PRACTICE BOOK

for

Conceptual Integrated Science

Third Edition

Paul G. Hewitt

Suzanne Lyons

John Suchocki

Jennifer Yeh

Senior Courseware Portfolio Manager: Jessica Moro
Director of Portfolio Management: Jeanne Zalesky
Courseware Editorial Assistant: Frances Lai, Sabrina Marshall
Managing Producer: Kristen Flathman
Full-Service Vendor: Pearson CSC
Cover Designer: Jeff Puda
Manufacturing Buyer: Stacey Weinberger
Director of Field Marketing: Tim Galligan
Director of Product Marketing: Allison Rona
Field Marketing Manager: Christopher Barker
Product Marketing Manger: Elizabeth Ellsworth Bell
Printer/Binder: LSC Communications, Inc.
Cover Photo Credit: age Dimtri Otis/Getty Images

ISBN 10: 0-13-547975-4; ISBN 13: 978-0-135-47975-9

---

3 2019

www.pearsonhighered.com

Table of Contents

To the Student

One of the things that people enjoy most is getting the best from their own brains—to discover we can comprehend concepts and ideas we thought we couldn't. This book of practice pages will help you to get the hang of science concepts. Understanding the major fields of science can be an enjoyable and worthwhile experience.

These pages are meant to *teach* you—not to *test* you. Use your textbook with these pages and consult with your classmates when you need help—or when they need help. Your teacher will probably not grade these, but will likely base your grade on test scores that indirectly relate to how well you do with these practice pages.

What you truly learn is yours, and can never be taken away from you.

To the Teacher

These are practice pages, NOT quiz sheets. They complement classroom discussion of chapter material. Whether used by students in small groups, in pairs, or even solo, they help your students "tie it all together." Please don't use these pages to grade your students. Ample test material can do that.

Please regard this book as your teaching tool. Get your students engaged in it and enjoy the results!

Name _____ Date _____

Conceptual Integrated Science — Third Edition

Chapter 1: About Science
Measuring the Size of Planet Earth

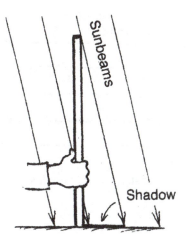

On a sunny day a stick held vertically with one end on the ground casts a shadow. In a region where a noon-time Sun is *directly overhead*, **no** shadow is cast.

With this common knowledge we can measure the roundness of Earth.

Shadow

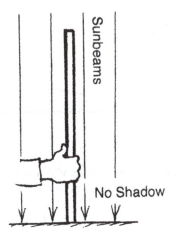

No Shadow

Consider a location on Earth where the Sun is directly overhead, where a vertical stick on the ground casts no shadow (lower stick in the diagram). At this special time, the angle between the Sun's rays and the stick is

(0°) (more than 0°) .

Circle the correct answer.

At the same moment a second vertical stick located 800 km north *does* cast a shadow. The length of this shadow is measured to be 1/8 the height of the stick.

Shadow 1/8 height of stick

Sunbeams

Earth

← 800 km →

Stick

Stick

If the stick were a meterstick, the length of the shadow would be

(1.25 cm)   (12.5 cm)   (80 cm).

If a line along each vertical stick were extended deep into Earth, it would pass through Earth's center. The lengths of these imaginary lines would be the same as Earth's

(radius) (diameter).

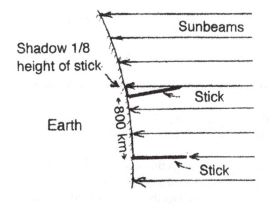

Shadow 1/8 height of stick

Sunbeams

Vertical to Earth's center

Vertical to Earth's center

Earth's center

Stick

Stick

Conceptual Integrated Science — Third Edition

Think ratio and proportion: There are two similar triangles to consider. First, a small one, two sides of which are the northern stick and its shadow. The stick is 8 times as long as its shadow.

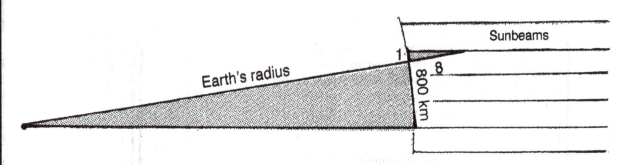

Second, a large triangle, whose corresponding sides are a vertical line that intersects Earth's center, labeled "Earth's radius," and the 800-km distance between the sticks at Earth's surface.

Equating the two ratios: $\dfrac{8}{1} = \dfrac{\text{Earth's radius}}{800\,\text{km}}$, and solving for Earth's radius we find

Earth's radius = (_____) × (_____km) = (_____km).

If your calculation for Earth's radius is 6400 kilometers, you are ready to calculate the circumference of Earth!

From Earth's radius, you know its circumference: $C = \pi D$, or, $2\pi R$.

Earth's circumference = $2\pi \times$ (_____km) = (40,000 km) (60,000 km) (80,000 km), remarkably within 5 percent of today's value.

The first to accomplish this was the Greek scholar and geographer Eratosthenes in about 235 BC. He didn't use a pair of sticks, but instead used a tall pillar in Alexandria and a deep well in the southern city of Syene, 800 kilometers distant. While the Sun was directly overhead in Syene, as evident by reflection of sunlight in the deep well, the pillar in Alexandria cast a shadow. Your textbook tells how he used the 7.2° angle between sunbeams and the pillar to calculate Earth's circumference.

> Quite interestingly, Earth's size can be calculated using *any* pair of vertical sticks a known distance apart if *at the same time* we know the angle between each stick and the sunbeams.

Vertical at Alexandria

Alexandria

Shadow

Vertical Pillar

Sunbeams

Earth's center

Syene

Vertical well

We remember Eratosthenes for his amazing calculation of Earth's size, using no tools but keen reasoning. Seventeen centuries after Eratosthenes' death, Christopher Columbus knew of Eratosthenes' findings before setting sail for what he thought would be China. Ignoring Eratosthenes, Columbus instead referred to more up-to-date maps that indicated a much smaller Earth. If he had relied on Eratosthenes' larger circumference, he would have known that he hadn't reached China, but instead had discovered a new world—later to be called America.

Name _____   Date _____

Conceptual Integrated Science — Third Edition

Chapter 2:   Describing Motion
Vectors and Equilibrium

Nellie Newton dangles from a vertical rope in equilibrium: $\sum F = 0$. The tension in the rope (upward vector) has the same magnitude as the downward pull of gravity (downward vector).

1. Nellie is supported by two vertical ropes. Draw tension vectors to scale along the direction of each rope.

2. This time the vertical ropes have different lengths. Draw tension vectors to scale for each of the two ropes.

3. Nellie is supported by three vertical ropes that are equally taut but have different lengths. Again, draw tension vectors to scale for each of the three ropes.

Circle the correct answer:

4. We see that tension in a rope is   (dependent on)   (independent of)   the length of the rope. So the length of a vector representing rope tension is   (dependent on)   (independent of)   the length of the rope.

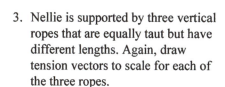

Rope tension depends on the angle the rope makes with the vertical, as Practice Pages for Chapter 3 will show!

Conceptual Integrated Science — Third Edition

Chapter 2:   Describing Motion

Free Fall Speed

1. Aunt Minnie gives you $10 per second for 4 seconds. How much money do you have after 4 seconds?
   _____

2. A ball dropped from rest picks up speed at 10 m/s per second. After it falls for 4 seconds, how fast is it going?
   _____

3. You have $20, and Uncle Harry gives you $10 each second for 3 seconds. How much money do you have after 3 seconds? _____

4. A ball is thrown straight down with an initial speed of 20 m/s. After 3 seconds, how fast is it going? _____

5. You have $50 and you pay Aunt Minnie $10/second. When will your money run out?
   _____

6. You shoot an arrow straight up at 50 m/s. When will it run out of speed? _____

7. What will be the arrow's speed 5 seconds after you shoot it? _____

8. What will its speed be 6 seconds after you shoot it? 7 seconds? _____
   _____

Free Fall Distance

1. Speed is one thing; distance another. *Where* is the arrow you shoot up at 50 m/s when it runs out of speed? _____

2. How high will the arrow be 7 seconds after being shot up at 50 m/s? _____

3. a. Aunt Minnie drops a penny into a wishing well and it falls for 3 seconds before hitting the water. How fast is it going when it hits? _____

   b. What is the penny's average speed during its 3 second drop? _____

   c. How far down is the water surface? _____

FROM REST,
$v = 10t$
$d = 5t^2$

4. Aunt Minnie didn't get her wish, so she goes to a deeper wishing well and throws a penny straight down into it at 10 m/s. How far does this penny go in 3 seconds? _____

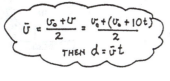

$$\bar{v} = \frac{v_0 + v}{2} = \frac{v_0 + (v_0 + 10t)}{2}$$

THEN $d = \bar{v}t$

Distinguish between " how fast,"
" how far," and " how long "!

Name _____ Date _____

Conceptual Integrated Science
Third Edition

Acceleration of Free Fall

A rock dropped from the top of a cliff picks up speed as it falls. Pretend that a speedometer and odometer are attached to the rock to show readings of speed and distance at 1-second intervals. Both speed and distance are zero at time = zero (see sketch). Note that after the rock falls 1 second the speed reading is 10 m/s and the distance fallen is 5 m. The readings for succeeding seconds of fall are not shown and are left for you to complete. Draw the position of the speedometer pointer and write in the correct odometer reading for each time. Use $g = 10 \text{ m/s}^2$ and neglect air resistance.

YOU NEED TO KNOW:
Instantaneous speed of fall from rest:
$$v = gt$$
Distance fallen from rest:
$$d = \tfrac{1}{2}gt^2$$

1. The speedometer reading increases by the same amount, _____ m/s, each second. This increase in speed per second is called _____.

2. The distance fallen increases as the square of the _____.

3. If it takes 7 seconds to reach the ground, then its speed at impact is _____ m/s, the total distance fallen is _____ m, and its acceleration of fall just before impact _____ m/s².

t = 0 s

t = 1 s

t = 2 s

t = 3 s

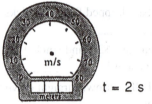
t = 4 s

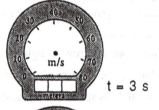
t = 5 s

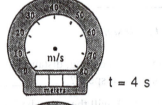
t = 6 s

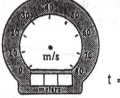

Conceptual Integrated Science Third Edition

Chapter 3: Newton's Laws of Motion

Newton's First Law and Friction

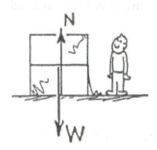

1. A crate filled with video games rests on a horizontal floor. Only gravity and the support force of the floor act on it, as shown by the vectors for weight **W** and normal force **N**.

   a. The net force on the crate is (zero) (greater than zero).

   b. Evidence for this is _____.

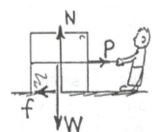

2. A slight pull **P** is exerted on the crate, not enough to move it. A force of friction **f** now acts,

   a. which is (less than) (equal to) (greater than) **P**.

   b. Net force on the crate is (zero) (greater than zero).

3. Pull **P** is increased until the crate begins to move. It is pulled so that it moves with constant velocity across the floor.

   a. Friction **f** is (less than) (equal to) (greater than) **P**.

   b. Constant velocity means acceleration is (zero) (greater than zero).

   c. Net force on the crate is (less than) (equal to) (greater than) zero.

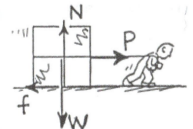

4. Pull **P** is further increased and is now greater than friction **f**.

   a. Net force on the crate is (less than) (equal to) (greater than) zero.

   b. The net force acts toward the right, so acceleration acts toward the (left) (right).

5. If the pulling force **P** is 150 N and the crate doesn't move, what is the magnitude of **f**? _____

6. If the pulling force **P** is 200 N and the crate doesn't move, what is the magnitude of **f**? _____

7. If the force of sliding friction is 250 N, what force is necessary to keep the crate sliding at constant velocity? _____

8. If the mass of the crate is 50 kg and sliding friction is 250 N, what is the acceleration of the crate when the pulling force is 250 N? _____ 300 N? _____ 500 N? _____

Name _____   Date _____

Conceptual Integrated Science — Third Edition

Nonaccelerated Motion

1. The sketch shows a ball rolling at constant velocity along a level floor. The ball rolls from the first position shown to the second in 1 second. The two positions are 1 meter apart. Sketch the ball at successive 1-second intervals all the way to the wall (neglect resistance).

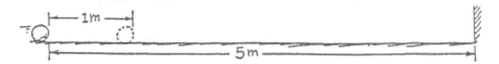

a. Did you draw successive ball positions evenly spaced, farther apart, or close together? Why?

_____

b. The ball reaches the wall with a speed of _____ m/s and takes a time of _____ seconds.

2. Table 1 shows the data of sprinting speeds of some animals. Make whatever computations are necessary to complete the table.

Table 1

ANIMAL	DISTANCE	TIME	SPEED
CHEETAH	75 m	3 s	25 m/s
GREYHOUND	160 m	10 s	
GAZELLE	1 km		100 km/h
TURTLE		30 s	1 cm/s

Accelerated Motion

3. An object starting from rest gains a speed $v = at$ when it undergoes uniform acceleration. The distance it covers is $d = 1/2\ at^2$. Uniform acceleration occurs for a ball rolling down an inclined plane. The plane below is tilted so a ball picks up a speed of 2 m/s each second; then its acceleration is $a = 2$ m/s². The positions of the ball are shown at 1-second intervals. Complete the six blank spaces for distance covered, and the four blank spaces for speeds.

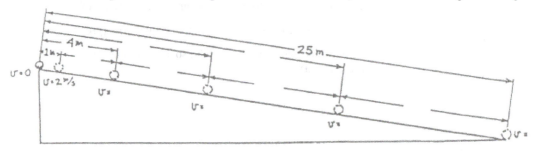

a. Do you see that the total distance from the starting point increases as the square of the time? This was $v = at$ as discovered by Galileo. If the incline were to continue, predict the ball's distance from the starting point for the next 3 seconds.

_____

b. Note the increase of distance between ball positions with time. Do you see an odd-integer pattern (also discovered by Galileo) for the increase? If the incline were to continue, predict the successive distances between ball positions for the next 3 seconds.

_____

Name _____   Date _____

Conceptual Integrated Science
Third Edition

Chapter 3:   Newton's Laws of Motion

A Day at the Races with Newton's Second Law: $a = \frac{F}{m}$

In each situation below, Cart A has a mass of **1 kg**. The mass of Cart B varies as indicated. Circle the correct answer (A, B, or Same for both).

1. Cart A is pulled with a force of **1 N**. Cart B also has a mass of **1 kg** and is pulled with a force of **2 N**. Which undergoes the greater acceleration?

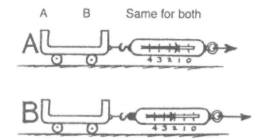

2. Cart A is pulled with a force of **1 N**. Cart B has a mass of **2 kg** and is also pulled with a force of **1 N**. Which undergoes the greater acceleration?

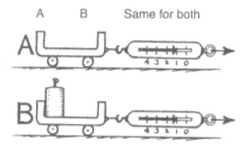

3. Cart A is pulled with a force of **1 N**. Cart B has a mass of **2 kg** and is pulled with a force of **2 N**. Which undergoes the greater acceleration?

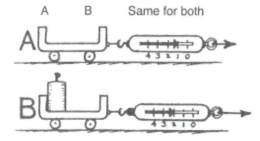

4. Cart A is pulled with a force of **1 N**. Cart B has a mass of **3 kg** and is pulled with a force of **3 N**. Which undergoes the greater acceleration?

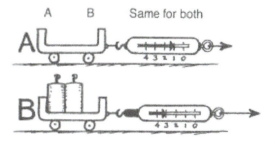

5. This time Cart A is pulled with a force of **4 N**. Cart B has a mass of **4 kg** and is pulled with a force of **4 N**. Which undergoes the greater acceleration?

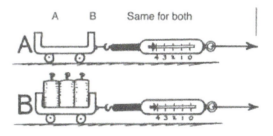

6. Cart A is pulled with a force of **2 N**. Cart B has a mass of **4 kg** and is pulled with a force of **3 N**. Which undergoes the greater acceleration?

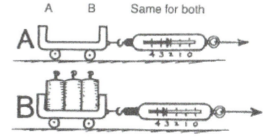

Conceptual Integrated Science ── Third Edition

Chapter 3:   Newton's Laws of Motion
Dropping Masses and Accelerating Cart

1. Consider the simple case of a 1-kg cart being pulled by a 10 N applied force. According to Newton's Second Law, acceleration of the cart is

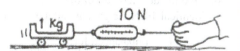

$$a = \frac{F}{m} = \frac{10\ N}{1\ kg} = 10\ m/s^2$$

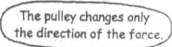
This is the same as the acceleration of free fall, *g*—because a force equal to the cart's weight accelerates it.

2. Now consider the acceleration of the cart when a second mass is also accelerated. This time the applied force is due to a 10-N iron weight attached to a string draped over a pulley. Will the cart accelerate as before, at 10 m/s² ? The answer is *no*, because the mass being accelerated is the mass of the cart *plus* the mass of the piece of iron that pulls it. Both masses accelerate. The mass of the 10-N iron weight is 1 kg, so the total mass being accelerated (cart + iron) is 2 kg. Then,

The pulley changes only the direction of the force.

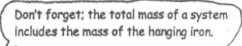

$$a = \frac{F}{m} = \frac{10\ N}{2\ kg} = 5\ m/s^2$$

Don't forget: the total mass of a system includes the mass of the hanging iron.

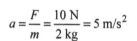
Note this is half the acceleration due to gravity alone, *g*. So the acceleration of 2 kg produced by the weight of 1 kg is *g*/2.

a. Find the acceleration of the 1-kg cart when two identical 10-N weights are attached to the string.

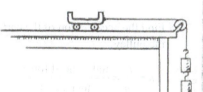

$$a = \frac{F}{m} = \frac{unbalanced\ force}{total\ mass} = \underline{\hspace{3cm}} = \underline{\hspace{1.5cm}}\ m/s^2.$$

Note that the mass being accelerated is 1 kg for the cart + 1 kg each for the weights = 3 kg.

Conceptual Integrated Science
Third Edition

Dropping Masses and Accelerating Cart—continued

b. Find the acceleration of the 1-kg cart when three identical 10-N weights are attached to the string.

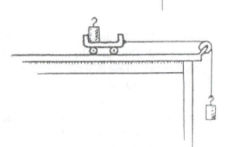

$$a = \frac{F}{m} = \frac{\text{unbalanced force}}{\text{total mass}} = \underline{\hspace{2cm}} = \underline{\hspace{1cm}} \text{ m/s}^2.$$

c. Find the acceleration of the 1-kg cart when four identical 10-N weights (not shown) are attached to the string.

$$a = \frac{F}{m} = \frac{\text{unbalanced force}}{\text{total mass}} = \underline{\hspace{2cm}} = \underline{\hspace{1cm}} \text{ m/s}^2.$$

d. This time, 1 kg of iron is added to the cart, and only one iron piece dangles from the pulley. Find the acceleration of the cart.

$$a = \frac{F}{m} = \frac{\text{unbalanced force}}{\text{total mass}} = \underline{\hspace{2cm}} = \underline{\hspace{1cm}} \text{ m/s}^2.$$

The force due to gravity on a mass m is mg.
So gravitational force on 1 kg is (1 kg)(10 m/s²) = 10 N.

e. Find the acceleration of the cart when it carries two pieces of iron and only one iron piece dangles from the pulley.

$$a = \frac{F}{m} = \frac{\text{unbalanced force}}{\text{total mass}} = \underline{\hspace{2cm}} = \underline{\hspace{1cm}} \text{ m/s}^2.$$

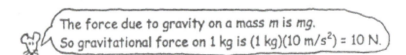

Conceptual Integrated Science Third Edition

Dropping Masses and Accelerating Cart—continued

f. Find the acceleration of the cart when it carries three pieces
    of iron and only one iron piece dangles from the pulley.

$$a = \frac{F}{m} = \frac{\text{unbalanced force}}{\text{total mass}} = \text{_____} = \text{_____} \text{ m/s}^2.$$

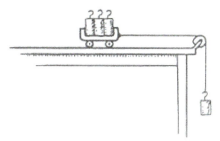

g. Find the acceleration of the cart when it carries three pieces of iron and
    four iron pieces dangle from the pulley.

$$a = \frac{F}{m} = \frac{\text{unbalanced force}}{\text{total mass}} = \text{_____} = \text{_____} \text{ m/s}^2.$$

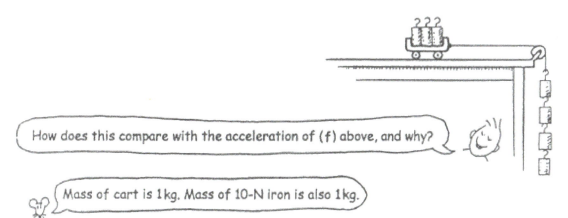

How does this compare with the acceleration of (f) above, and why?

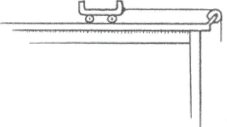

Mass of cart is 1 kg. Mass of 10-N iron is also 1 kg.

h. Draw your own combination of masses and find the acceleration.

$$a = \frac{F}{m} = \frac{\text{unbalanced force}}{\text{total mass}} = \text{_____} = \text{_____} \text{ m/s}^2.$$

Conceptual Integrated Science
Third Edition

Chapter 3: Newton's Laws of Motion
Mass and Weight

Learning physics is learning the connections among concepts in nature, and also learning to distinguish between closely related concepts. Velocity and acceleration are often confused. Similarly, in this chapter, we find that mass and weight are often confused. They aren't the same! Please review the distinction between mass and weight in your textbook. To reinforce your understanding of this distinction, circle the correct answers below.

Comparing the concepts of mass and weight, one is basic—fundamental—depending only on the internal makeup of an object and the number and kind of atoms that compose it. The concept that is fundamental is   (mass)   (weight).

The concept that additionally depends on location in a gravitational field is   (mass)   (weight).

To repeat for emphasis   (Mass)   (Weight)   is a measure of the amount of matter in an object and only depends on the number and kind of atoms that compose it.

We can correctly say that   (mass)   (weight)   is a measure of an object's "laziness."

(Mass)   (Weight)   is related to the gravitational force acting on the object.

(Mass)   (Weight)   depends on an object's location, whereas   (mass)   (weight)   does not.

In other words, a stone would have the same   (mass)   (weight)   whether it is on Earth's surface or the Moon's surface. However, its   (mass)   (weight)   depends on its location.

On the Moon's surface, where gravity is only about 1/16 of Earth's gravity,   (mass)   (weight)   (both the mass and the weight)   of the stone would be the same as on Earth.

While mass and weight are not the same, they are   (directly proportional)   (inversely proportional)   to each other.

In the same location, twice the mass has   (twice)   (half)   the weight.

The Standard International (SI) unit of mass is the   (kilogram)   (newton), and the SI unit of force is the   (kilogram)   (newton).

In the United States, it is common to measure the mass of something by measuring its gravitational pull to Earth, its weight. The common unit of weight in the United States is the   (pound)   (kilogram)   (newton).

Pull of gravity

Support Force

When I step on a scale, two forces act on it: a downward pull of gravity, and an upward support force. These equal and opposite forces effectively compress a spring inside the scale that is calibrated to show weight. When in equilibrium, my weight = *mg*.

Conceptual Integrated Science — Third Edition

Chapter 3:   Newton's Laws of Motion

Converting Mass to Weight

Objects with mass also have weight (although they can be weightless under special conditions). If you know the mass of something in **kilograms** and want its weight in **newtons**, at Earth's surface, you can take advantage of the formula that relates weight and mass:

$$\text{Weight} = \text{mass} \times \text{acceleration due to gravity}$$
$$W = mg.$$

This is in accord with Newton's Second Law, written as $F = ma$. When the force of gravity is the only force, the acceleration of any object of mass m will be g, the acceleration of free fall. Importantly, g acts as a proportionality constant, 9.8 N/kg, which is equivalent to 9.8 m/s^2.

Sample Question:

How much does a 1-kg bag of nails weigh on Earth?

$W = mg = (1 \text{ kg})(9.8 \text{ m/s}^2) = 9.8 \text{ m/s}^2 = 9.8 \text{ N}.$

or simply, $W = mg = (1 \text{ kg})(9.8 \text{ N/kg}) = 9.8 \text{ N}.$

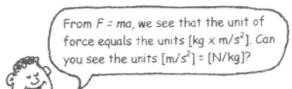

From $F = ma$, we see that the unit of force equals the units [kg × m/s²]. Can you see the units [m/s²] = [N/kg]?

Answer the following questions:

Felicia the ballet dancer has a mass of 45 kg.

1. What is Felicia's weight in newtons on Earth's surface? _____

2. Given that 1 kilogram of mass corresponds to 2.2 pounds on Earth's surface, what is Felicia's weight in pounds on Earth? _____

3. What would be Felicia's mass on the surface of Jupiter? _____

4. What would be Felcia's weight on Jupiter's surface, where the acceleration due to gravity is 25.0 m/s^2?

   _____

Different masses are hung on a spring scale calibrated in newtons. The force exerted by gravity on $1 \text{ kg} = 9.8 \text{ N}$.

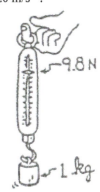

5. The force exerted by gravity on 5 kg = _____ N.

6. The force exerted by gravity on _____ kg = 98 N.

Make up your own mass and show the corresponding weight:

The force exerted by gravity on _____ kg = _____ N.

By whatever means (spring scales, measuring balances, etc.), find the mass of your integrated science book. Then complete the table.

OBJECT	MASS	WEIGHT
MELON	1 kg	
APPLE		1 N
BOOK		
A FRIEND	60 kg	

Chapter 3:   Newton's Laws of Motion

Bronco and Newton's Second Law

Bronco skydives and parachutes from a stationary helicopter. Various stages of fall are shown in positions *a* through *f*. Using Newton's Second Law

$$a = \frac{F_{NET}}{m} = \frac{W - R}{m}$$

find Bronco's acceleration at each position (answer in the blanks to the right). You need to know that Bronco's mass *m* is 100 kg so his weight is a constant 1000 N. Air resistance *R* varies with speed and cross-sectional area as shown.

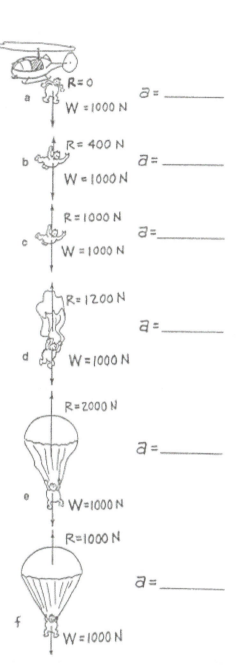

R = 0
a
W = 1000 N
$a = $ _____

R = 400 N
b
W = 1000 N
$a = $ _____

R = 1000 N
c
W = 1000 N
$a = $ _____

R = 1200 N
d
W = 1000 N
$a = $ _____

R = 2000 N
e
W = 1000 N
$a = $ _____

R = 1000 N
f
W = 1000 N
$a = $ _____

Circle the correct answers:

1. When Bronco's speed is least, his acceleration is

   (least)   (most).

2. In which position(s) does Bronco experience a downward acceleration?

   (a)   (b)   (c)   (d)   (e)   (f)

3. In which position(s) does Bronco experience an upward acceleration?

   (a)   (b)   (c)   (d)   (e)   (f)

4. When Bronco experiences an upward acceleration, his velocity is

   (still downward)   (upward also).

5. In which position(s) is Bronco's velocity constant?

   (a)   (b)   (c)   (d)   (e)   (f)

6. In which position(s) does Bronco experience terminal velocity?

   (a)   (b)   (c)   (d)   (e)   (f)

7. In which position(s) is terminal velocity greatest?

   (a)   (b)   (c)   (d)   (e)   (f)

8. If Bronco were heavier, his terminal velocity would be

   (greater)   (less)   (the same).

Conceptual Integrated Science
Third Edition

Chapter 3: Newton's Laws of Motion
Newton's Third Law

Your thumb and finger pull on each other when you stretch a rubber band between them. This pair of forces, thumb on finger and finger on thumb, make up an action–reaction pair of forces, both of which are equal in magnitude and oppositely directed. Draw the reaction vector and state in words the reaction force for each of the examples **a** through **g**. Then make up your own example in **h**.

Thumb pulls finger

Finger pulls thumb

Foot hits ball

a _____

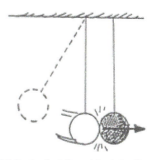

White ball strikes black ball

b _____

Earth pulls on the Moon

c _____

Tires push backward on road

d _____

Wings push air downward

e _____

Fish pushes water backward

f _____

Helen touches Hyrum

g _____

h _____

YOU CAN'T TOUCH
WITHOUT BEING TOUCHED—
NEWTON'S THIRD LAW

Name _____ Date _____

Conceptual Integrated Science
Third Edition

Chapter 3: Newton's Laws of Motion
Nellie and Newton's Third Law

Nellie holds an apple weighing 1 newton at rest on the palm of her hand. *Circle the correct answers.*

1. To say the weight (W) of the apple is 1 N is to say that a downward gravitational force of 1 N is exerted on the apple by

   (Earth)   (her hand).

2. Nellie's hand supports the apple with normal force N, which acts in a direction opposite to W. We can say N

   (equals W)   (has the same magnitude as W).

3. Since the apple is at rest, the net force on the apple is

   (zero)   (nonzero).

4. Since N is equal and opposite to W, we (can) (cannot) say that N and W constitute an action–reaction pair. The reason is that action

   and reaction   (act on the same object)   (act on different objects), and here we see N and W (both acting on the apple)   (acting on different objects).

5. In accord with the rule "If ACTION is A acting on B, then REACTION is B acting on A," if we say action is Earth pulling down on the apple, reaction is

   (the apple pulling up on Earth)   (N, Nellie's hand pushing up on the apple).

6. To repeat for emphasis, we see that N and W are equal and opposite to each other

   (and constitute an action–reaction pair)   (but do *not* constitute an action–reaction pair).

To identify a pair of action–reaction forces in any situation, first identify the pair of interacting objects involved. Something is interacting with something else. In this case, the whole Earth is interacting (gravitationally) with the apple. So, Earth pulls downward on the apple (call it action), while the apple pulls upward on Earth (reaction).

Simply put, Earth pulls on apple (action); apple pulls on Earth (reaction).

Better put, apple and Earth *pull on each other* with equal and opposite forces that constitute a *single* interaction.

7. Another pair of forces is N [shown] and the downward force of the apple against Nellie's hand [not shown].
   This pair of forces   (is)   (isn't)   an action–reaction pair.

8. Suppose Nellie now pushes upward on the apple with the force of 2 N. The apple   (is still in equilibrium) (accelerates upward), and compared with W, the magnitude of N is   (the same)   (twice) (not the same, and not twice).

9. Once the apple leaves Nellie's hand, N is   (zero)   (still twice the magnitude of W), and the net force on the apple is   (zero)   (only W)   (still W – N, which is a negative force).

Conceptual Integrated Science — Third Edition

Chapter 3: Newton's Laws of Motion

Vectors and the Parallelogram Rule

1. When vectors **A** and **B** are at an angle to each other, they add to produce the resultant **C** by the *parallelogram rule*. Note that **C** is the diagonal of a parallelogram where **A** and **B** are adjacent sides. Resultant **C** is shown in the first two diagrams, *a* and *b*. Construct the resultant **C** in diagrams *c* and *d*. Note that in diagram *d* you form a rectangle (a special case of a parallelogram).

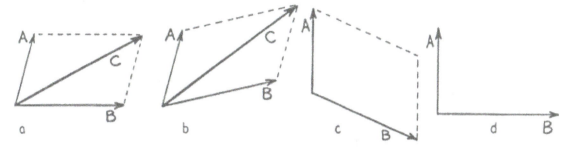

2. Below we see a top view of an airplane being blown off course by wind in various directions. Use the parallelogram rule to show the resulting speed and direction of travel for each case. In which case does the airplane travel fastest across the ground? _____ Slowest? _____

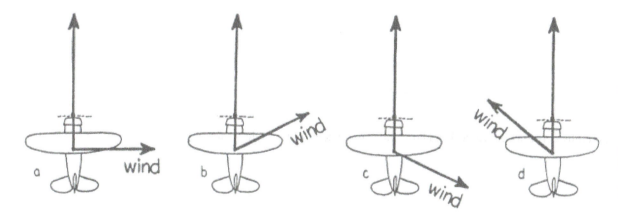

3. To the right we see top views of three motorboats crossing a river. All have the same speed relative to the water, and all experience the same water flow.

   Construct resultant vectors showing the speed and direction of the boats.

   a. Which boat takes the shortest path to the opposite shore? _____

   b. Which boat reaches the opposite shore first? _____

   c. Which boat provides the fastest ride? _____

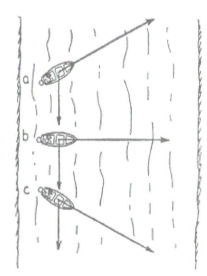

Conceptual Integrated Science
Third Edition

Vectors

Use the parallelogram rule to carefully construct the resultants for the eight pairs of vectors.

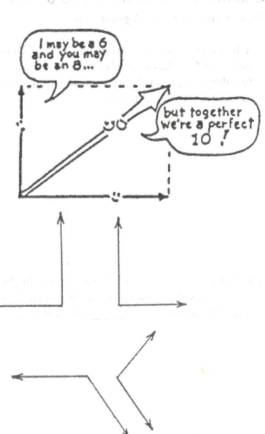

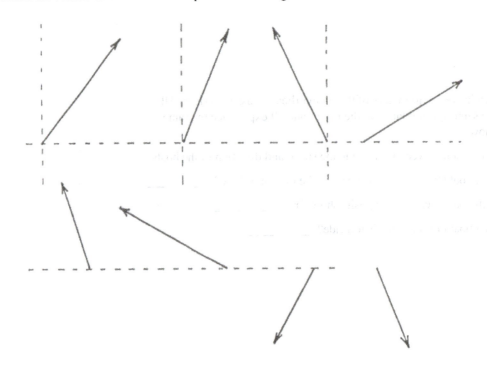

Carefully construct the vertical and horizontal components of the eight vectors.

Conceptual Integrated Science
Third Edition

Chapter 3: Newton's Laws of Motion
Force Vectors and the Parallelogram Rule

1. The heavy ball is supported in each case by two strands of rope. The tension in each strand is shown by the vectors. Use the parallelogram rule to find the resultant of each vector pair.

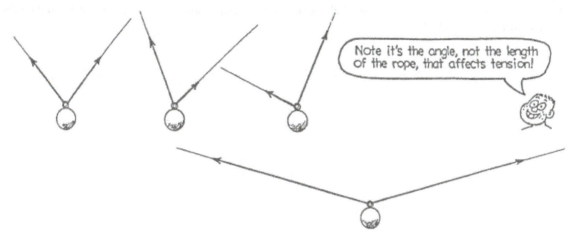

Note it's the angle, not the length of the rope, that affects tension!

a. Is your resultant vector the same for each case? _____

b. How do you think the resultant vector compares to the weight of the ball?

_____

2. Now let's do the opposite of what we've done above. More often, we know the weight of the suspended object, but we don't know the rope tensions. In each case below, the weight of the ball is shown by the vector W. Each dashed vector represents the resultant of the pair of rope tensions. Note that each is equal and opposite to vector W (they must be; otherwise the ball wouldn't be at rest).

a. Construct parallelograms where the ropes define adjacent sides and the dashed vectors are the diagonals.

b. How do the relative lengths of the sides of each parallelogram compare to rope tensions?

c. Draw rope-tension vectors, clearly showing their relative magnitudes.

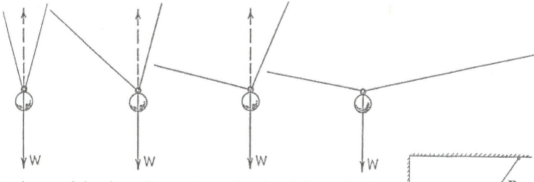

3. A lantern is suspended as shown. Draw vectors to show the relative tensions in ropes **A, B,** and **C.** Do you see a relationship between your vectors **A + B** and vector **C?** Between vectors **A + C** and vector **B?**

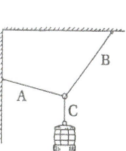

Conceptual Integrated Science — **Third Edition**

Force-Vector Diagrams

In each case, a rock is acted on by one or more forces. Draw an accurate vector diagram showing all forces acting on the rock, and no other forces. Use a ruler, and do it in pencil so you can correct mistakes. The first two are done as examples. Show by the parallelogram rule in 2 that the vector sum of **A** + **B** is equal and opposite to **W** (i.e., **A** + **B** = − **W**). Do the same for 3 and 4. Draw and label vectors for the weight and normal forces in 5 to 10, and for the appropriate forces in 11 and 12.

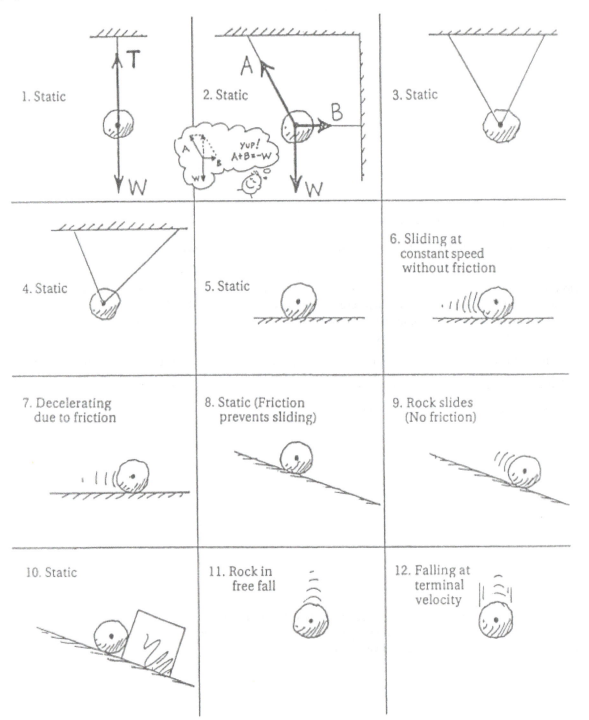

1. Static

2. Static   yup! A+B=−w

3. Static

4. Static

5. Static

6. Sliding at constant speed without friction

7. Decelerating due to friction

8. Static (Friction prevents sliding)

9. Rock slides (No friction)

10. Static

11. Rock in free fall

12. Falling at terminal velocity

Name _____ Date _____

Conceptual Integrated Science
Third Edition

Chapter 4:   Momentum and Energy

Momentum

1. A moving car has momentum. If it moves twice as fast, its momentum is _____ as much.

2. Two cars, one twice as heavy as the other, move down a hill at the same speed. Compared with the lighter car, the momentum of the heavier car is _____ as much.

3. The recoil momentum of a gun that kicks is

   (more than)   (less than)   (the same as)

   the momentum of the gases and bullet it fires.

4. If a man firmly holds a gun when fired, then the momentum of the bullet and expelled gases is equal to the recoil momentum of the

   (gun alone)   (gun–man system)   (man alone).

5. Suppose you are traveling in a bus at highway speed on a nice summer day and the momentum of an unlucky bug is suddenly changed as it splatters onto the front window.

   a. Compared to the force that acts on the bug, how much force acts on the bus?

      (more)   (the same)   (less)

   b. The time of impact is the same for both the bug and the bus. Compared with the impulse on the bug, this means the impulse on the bus is

      (more)   (the same)   (less).

   c. Although the momentum of the bus is very large compared with the momentum of the bug, the change in momentum of the bus compared with the *change* of momentum of the bug is

      (more)   (the same)   (less).

   d. Which undergoes the greater acceleration?

      (bus)   (both the same)   (bug)

   e. Which, therefore, suffers the greater damage?

      (bus)   (both the same)   (The bug, of course!)

Conceptual Integrated Science
Third Edition

Chapter 4: Momentum and Energy

Systems

Momentum conservation (and Newton's Third Law) applies to *systems* of bodies. Here we identify some systems.

1. When the compressed spring is released, Blocks A and B will slide apart. There are three systems to consider here, indicated by the closed dashed lines below—System A, System B, and System A + B. Ignore the vertical forces of gravity and the support force of the table.

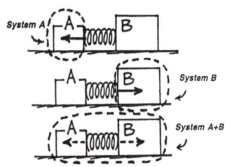

a. Does an external force act on System A?   (yes)   (no)

   Will the momentum of System A change?   (yes)   (no)

b. Does an external force act on System B?   (yes)   (no)

   Will the momentum of System B change?   (yes)   (no)

c. Does an external force act on System A + B?   (yes)   (no)

   Will the momentum of System A + B change?   (yes)   (no)

2. Billiard ball A collides with billiard ball B at rest. Isolate each system with a closed dashed line. Draw only the external force vectors that act on each system.

      System A   System B   System A+B

a. Upon collision, the momentum of System A   (increases)   (decreases)   (remains unchanged).

b. Upon collision, the momentum of System B   (increases)   (decreases)   (remains unchanged).

c. Upon collision, the momentum of System A + B   (increases)   (decreases)   (remains unchanged).

3. A girl jumps upward from Earth's surface. In the sketch to the left, draw a closed dashed line to indicate the system of the girl.

   a. Is there an external force acting on her?   (yes)   (no)

      Does her momentum change?   (yes)   (no)

      Is the girl's momentum conserved?   (yes)   (no)

   b. In the sketch to the right, draw a closed dashed line to indicate the system (girl + Earth). Is there an external force due to the interaction between the girl and Earth that acts on the system?   (yes)   (no)

      Is the momentum of the system conserved?   (yes)   (no)

4. A block strikes a blob of jelly. Isolate three systems with a closed dashed line and show the external force on each. In which system is momentum conserved?

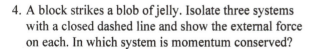

5. A truck crashes into a wall. Isolate three systems with a closed dashed line and show the external force on each. In which system is momentum conserved?

Name _____ Date _____

Conceptual Integrated Science — Third Edition

Chapter 4: Momentum and Energy

Impulse–Momentum

Bronco Brown wants to put $Ft = \Delta mv$ to the test and try bungee jumping. Bronco leaps from a high cliff and experiences free fall for 3 seconds. Then the bungee cord begins to stretch, reducing his speed to zero in 2 seconds. Fortunately, the cord stretches to its maximum length just short of the ground below.

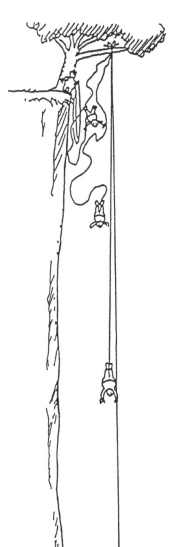

$t = 0$ s $v = $ _____

     momentum = _____

$t = 1$ s $v = $ _____

     momentum = _____

$t = 2$ s $v = $ _____

     momentum = _____

$t = 3$ s $v = $ _____

     momentum = _____

$t = 5$ s $v = $ _____

     momentum = _____

Fill in the blanks. Bronco's mass is 100 kg. Acceleration of free fall is 10 m/s^2.

Express values in SI units (distance in m, velocity in m/s, momentum in kg · m/s, impulse in N · s, and deceleration in m/s^2).

1. The 3-s free-fall distance of Bronco just before the bungee cord begins to stretch = _____

2. Δmv during the 3-s interval of free fall = _____

3. Δmv during the 2-s interval of slowing down = _____

4. *Impulse* during the 2-s interval of slowing down = _____

5. *Average force* exerted by the cord during the 2-s interval of slowing down = _____

6. How about *work* and *energy*? How much KE does Bronco have 3 s after his jump? _____

7. How much does gravitational PE decrease during this 3 s? _____

8. What two kinds of PE are changing during the slowing-down interval?

_____

Conceptual Integrated Science
Third Edition

Chapter 4:   Momentum and Energy

Conservation of Momentum

Granny whizzes around the rink and is suddenly confronted with Ambrose at rest directly in her path. Rather than knock him over, she picks him up and continues in motion without "braking." Consider both Granny and Ambrose as two parts of one system. Since no outside forces act on the system, the momentum of the system before collision equals the momentum of the system after collision.

a. Complete the before-collision data in the table below.

BEFORE COLLISION	
Granny's mass	80 kg
Granny's speed	3 m/s
Granny's momentum	_____
Ambrose's mass	40 kg
Ambrose's speed	0 m/s
Ambrose's momentum	__0__
Total momentum	_____

b. After collision, does Granny's speed increase or decrease?

_____

c. After collision, does Ambrose's speed increase or decrease?

_____

d. After collision, what is the total mass of Granny + Ambrose?

_____

e. After collision, what is the total momentum of Granny + Ambrose?

_____

f. Use the conservation of momentum law to find the speed of Granny and Ambrose together after collision. (Show your work in the space below.)

New speed =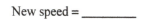

Conceptual Integrated Science
Third Edition

Chapter 4: Momentum and Energy
Work and Energy

1. How much work (energy) is needed to lift an object that weighs 200 N to a height of 4 m?

_____

2. How much power is needed to lift the 200-N object to a height of 4 m in 4 s?

_____

3. What is the power output of an engine that does 60,000 J of work in 10 s?

_____

4. The block of ice weighs 500 newtons.

   a. Neglecting friction, how much force is needed to push it up the incline?

_____

   b. How much work is required to push it up the incline compared with lifting the block vertically 3 m?

_____

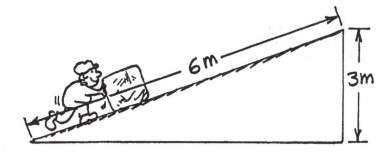

5. All the ramps are 5 m high. We know that the KE of the block at the bottom of the ramp will be equal to the loss of PE (conservation of energy). Find the speed of the block at ground level in each case. [Hint: Do you recall from earlier chapters how long it takes something to fall a vertical distance of 5 m from a position of rest (assume g = 10 m/s²)? And how much speed a falling object acquires in this time? This gives you the answer to Case 1. Discuss with your classmates how energy conservation gives you the answers to Cases 2 and 3.]

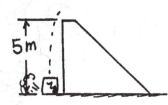  

Case 1: Speed = _____ m/s   Case 2: Speed = _____ m/s   Case 3: Speed = _____ m/s

Conceptual Integrated Science — Third Edition

Work and Energy—continued

6. Which block gets to the bottom of the incline first? Assume there is no friction. (Be careful!) Explain your answer.

   _____

   _____

   _____

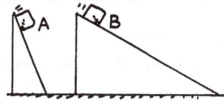

7. The KE and PE of a block freely sliding down a ramp are shown in only one place in the sketch. Fill in the missing information.

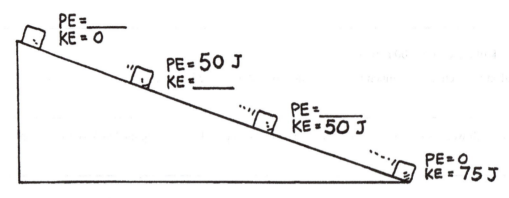

PE = _____
KE = 0

PE = 50 J
KE = _____

PE = _____
KE = 50 J

PE = 0
KE = 75 J

8. A big metal bead slides due to gravity along an upright friction-free wire. It starts from rest at the top of the wire as shown in the sketch. How fast is it traveling as it passes

   Point B? _____

   Point D? _____

   Point E? _____

   At what point does it have the maximum speed? _____

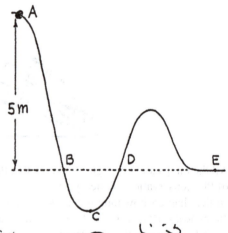

5 m

9. Rows of wind-powered generators are used in various windy locations to generate electric power. Does the power generated affect the speed of the wind? Would locations behind the "windmills" be windier if they weren't there? Discuss this in terms of energy conservation with your classmates.

   _____

   _____

   _____

   _____

Conceptual Integrated Science — Third Edition

Chapter 4: Momentum and Energy

Conservation of Energy

Fill in the blanks for the six systems shown:

υ = 30 km/h
KE = 10⁶ J

υ = 60 km/h
KE = _____

υ = 90 km/h
KE = _____

PE = 15 000 J
KE = 0

PE = 11250 J
KE = _____

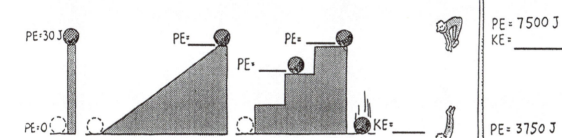

PE = 30 J

PE = _____

PE = _____

PE = _____

PE = 0

KE = _____

PE = 7500 J
KE = _____

PE = 3750 J
KE = _____

PE = 0 J
KE = _____

PE = 10⁴ J

WORK DONE = _____

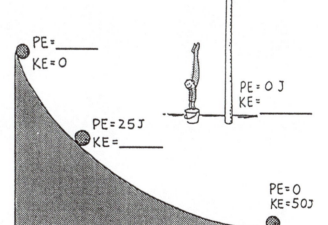

PE = _____
KE = 0

PE = 25 J
KE = _____

PE = 0
KE = 50 J

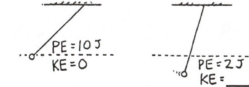

PE = 10 J
KE = 0

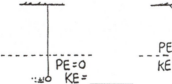

PE = 2 J
KE = _____

PE = 0
KE = _____

PE = _____
KE = _____

Conceptual Integrated Science
Third Edition

Chapter 4:   Momentum and Energy
Momentum, Impulse, and Kinetic Energy

A Honda Civic and a Lincoln Town Car are initially at rest in a horizontal parking lot at the edge of a steep cliff. For simplicity, we assume that the Town Car has twice as much mass as the Civic. Equal constant forces are applied to each car and they accelerate across equal distances (we ignore the effects of friction). When they reach the far end of the lot the force is suddenly removed, whereupon they sail through the air and crash to the ground below. (The cars are beat up to begin with, and this is a scientific experiment!)

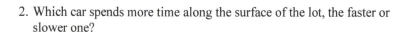

Let equations guide your thinking!

1. Which car has the greater acceleration? (Think $a = \frac{F}{m}$)

   _____

2. Which car spends more time along the surface of the lot, the faster or slower one?

   _____

3. Which car has the larger impulse imparted to it by the applied force? (Think Impulse = Ft.) Defend your answer.

   _____

4. Which car has the greater momentum at the cliff's edge? (Think $Ft = \Delta mv$.) Defend your answer.

Impulse = Δ momentum
$Ft = \Delta mv$

Work = Fd = ΔKE = $\Delta \frac{1}{2}mv^2$

   _____

5. Which car has the greater work done on it by the applied force? (Think $W = Fd$)
   Defend your answer in terms of the distance traveled.

   _____

6. Which car has the greater kinetic energy at the edge of the cliff? (Think $W = \Delta KE$)
   Does your answer follow from your explanation of Question 5? Does it contradict your answer to Question 3? Why or why not?

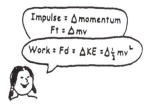

Making the distinction between momentum and kinetic energy is high-level physics.

   _____

7. Which car spends more time in the air, from the edge of the cliff to the ground below?

   _____

8. Which car lands farthest horizontally from the edge of the cliff onto the ground below?

   _____

Challenge: Suppose the slower car crashes a horizontal distance of 10 m from the ledge. At what horizontal distance does the faster car hit?

Conceptual Integrated Science — **Third Edition**

Chapter 5: Gravity

The Inverse-Square Law—Weight

1. Paint spray travels radially away from the nozzle of the can in straight lines. Like gravity, the strength (intensity) of the spray obeys an inverse-square law. Complete the diagram by filling in the blank spaces.

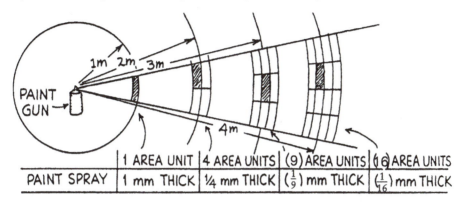

PAINT SPRAY	1 AREA UNIT	4 AREA UNITS	(9) AREA UNITS	(16) AREA UNITS
	1 mm THICK	¼ mm THICK	$(\frac{1}{9})$ mm THICK	$(\frac{1}{16})$ mm THICK

2. A small light source located 1 m in front of an opening of area $1\ m^2$ illuminates a wall behind. If the wall is 1 m behind the opening (2 m from the light source), the illuminated area covers $4\ m^2$. How many square meters will be illuminated if the wall is

   5 m from the source? _____

   10 m from the source? _____

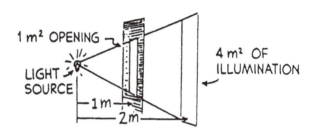

3. If we stand on a weighing scale and find that we are pulled toward Earth with a force of 500 N, then we weigh _____ N. Strictly speaking, we weigh _____ N relative to Earth. How much does Earth weigh? If we tip the scale upside down and repeat the weighing process, we can say that we and Earth are still pulled together with a force of _____ N, and, therefore, relative to us, the whole 6,000,000,000,000,000,000,000,000-kg Earth weighs _____ N! Weight, unlike mass, is a relative quantity.

We are pulled to Earth with a
force of 500 N, so we weigh 500 N.

Earth is pulled toward us with a
force of 500 N, so it weighs 500 N.

Conceptual Integrated Science
Third Edition

Chapter 5: Gravity
Ocean Tides

1. Consider two equal-mass blobs of water, A and B, initially at rest in the Moon's gravitational field. The vector shows the gravitational force of the Moon on A.

   a. Draw a force vector on B due to the Moon's gravity.

   b. Is the force on B more or less than the force on A? _____

   c. Why? _____

   d. The blobs accelerate toward the Moon. Which has the greater acceleration? (A)　(B)

   e. Because of the different accelerations, with time

      (A gets farther ahead of B)　(A and B gain identical speeds)　and the distance between A and B

      (increases)　(stays the same)　(decreases).

   f. If A and B were connected by a rubber band, with time the rubber band would

      (stretch)　(not stretch).

   g. This　(stretching)　(nonstretching)　is due to the　(difference)　(nondifference) in the Moon's gravitational pulls.

   h. The two blobs will eventually crash into the Moon. To orbit around the Moon instead of crashing into it, the blobs should move　(away from the Moon)　(tangentially). Then their accelerations will consist of changes in (speed)　(direction).

2. Now consider the same two blobs located on opposite sides of Earth.

   a. Because of differences in the Moon's pull on the blobs, they tend to

      (spread away from each other)　(approach each other). This produces ocean tides!

   b. If Earth and the Moon were closer, gravitational force between them would be

      (more)　(the same)　(less), and the difference in gravitational forces on the near and far parts of the ocean

      would be　(more)　(the same)　(less).

   c. Because Earth's orbit about the Sun is slightly elliptical, Earth and the Sun are closer in December than in June. Taking the Sun's tidal force into account, on a world average, ocean tides are greater in

      (December)　(June)　(no difference).

Conceptual Integrated Science — Third Edition

Chapter 5: Gravity
Projectile Motion

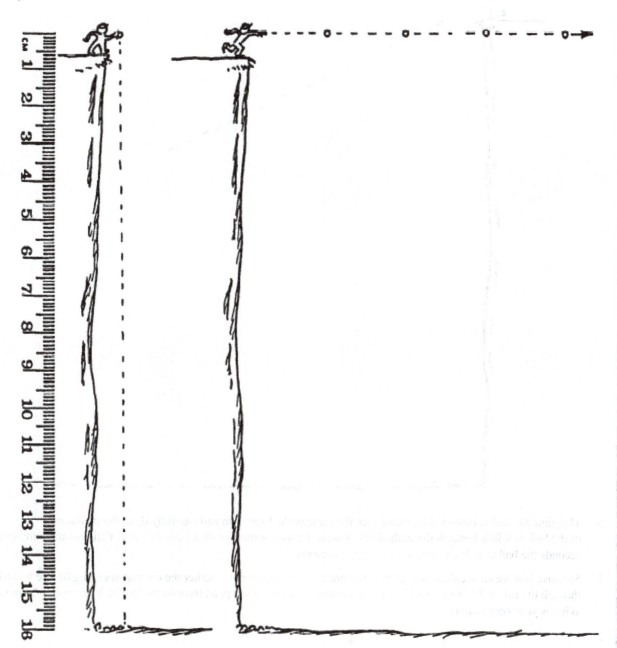

1. Above left: Use the scale 1 cm:5 m and draw the positions of the dropped ball at 1-second intervals. Neglect air drag and assume $g = 10 \text{ m/s}^2$. Estimate the number of seconds the ball is in the air.

   _____ seconds.

2. Above right: The four positions of the thrown ball with *no gravity* are at 1-second intervals. At 1 cm:5 m, carefully draw the positions of the ball *with* gravity. Neglect air drag and assume $g = 10 \text{ m/s}^2$. Connect your positions with a smooth curve to show the path of the ball. How is the motion in the vertical direction affected by motion in the horizontal direction?

Projectile Motion—continued

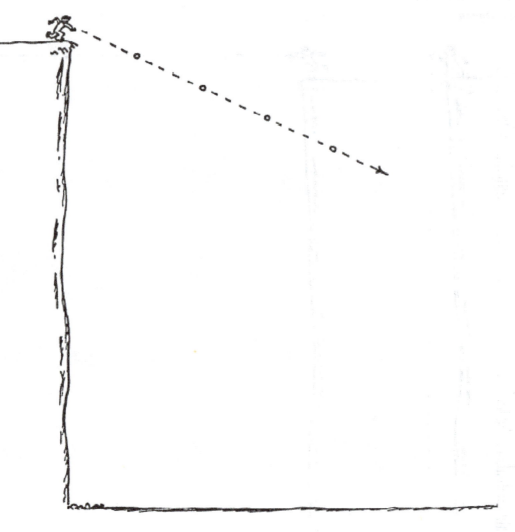

3. This time the ball is thrown downward. Use the same scale 1 cm, 5 m and carefully draw the positions
    of the ball as it falls beneath the dashed line. Connect your positions with a smooth curve. Estimate the number of
    seconds the ball remains in the air. _____ seconds

4. Suppose you are an accident investigator and are asked to figure out whether the car was speeding before it crashed
    through the rail of the bridge and into the mudbank as shown. The speed limit on the bridge is 55 mph = 24 m/s.
    What is your conclusion?

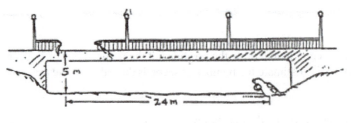

_____

_____

_____

Conceptual Integrated Science
Third Edition

Chapter 5: Gravity
Tossed-Ball Vectors

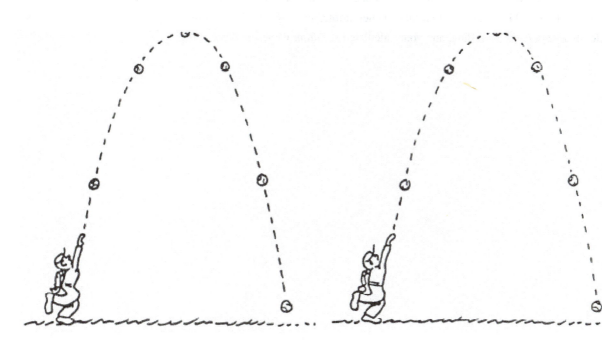

1. Draw sample vectors to represent the force of gravity on the ball in the positions shown above (after it leaves the thrower's hand). Neglect air drag.

2. Draw sample bold vectors to represent the velocity of the ball in the positions shown above. With lighter vectors, show the horizontal and vertical components of velocity for each position.

3. a. Which velocity component in the previous question remains constant? Why?

_____

b. Which velocity component changes along the path? Why?

_____

4. It is important to distinguish between force and velocity vectors. Force vectors combine with other force vectors, and velocity vectors combine with other velocity vectors. Do velocity vectors combine with force vectors?

_____

Conceptual Integrated Science — Third Edition

Tossed-Ball Vectors—continued

A ball tossed upward has initial velocity components 30 m/s vertical and 5 m/s horizontal. The position of the ball is shown at 1-second intervals. Air resistance is negligible, and $g = 10$ m/s^2. Fill in the boxes, writing in the values of velocity *components* ascending, and your calculated *resultant velocities* descending.

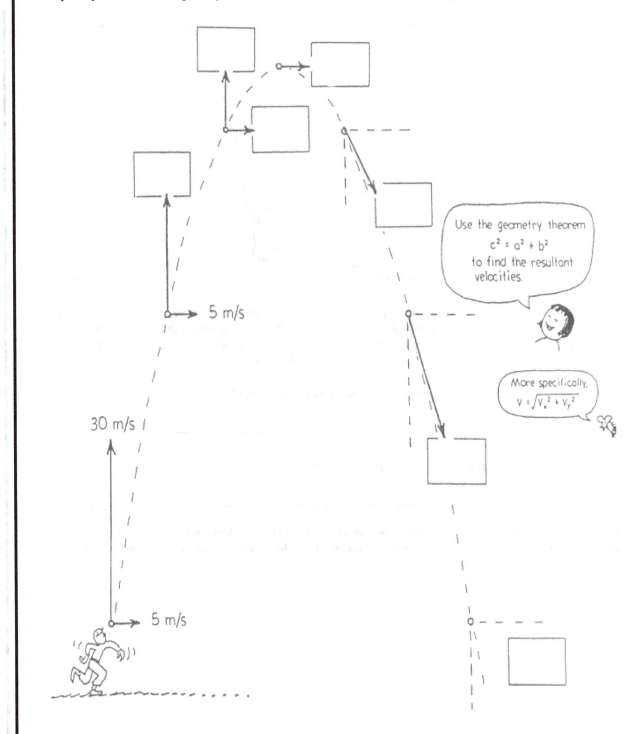

Use the geometry theorem
$$c^2 = a^2 + b^2$$
to find the resultant velocities.

More specifically,
$$V = \sqrt{V_x^2 + V_y^2}$$

Conceptual Integrated Science ─ Third Edition

Chapter 5: Gravity
Circular and Elliptical Orbits

I. Circular Orbits

1. Figure 1 shows "Newton's Mountain," so high that its top is above the drag of the atmosphere. The cannonball is fired and hits the ground as shown.

   a. Draw the path the cannonball might take if it were fired a little bit faster.

   b. Repeat for a still greater speed, but still less than 8 km/s.

   c. Draw the orbital path it would take if its speed were 8 km/s.

   d. What is the shape of the 8-km/s curve?

   e. What would be the shape of the orbital path if the cannonball were fired at a speed of about 9 km/s?

   Figure 1

2. Figure 2 shows a satellite in circular orbit.

   a. At each of the four positions draw a vector that represents the gravitational *force* exerted on the satellite.

   b. Label the force vectors **F**.

   c. Draw at each position a vector to represent the *velocity* of the satellite at that position and label it **V**.

   d. Are all four **F** vectors the same length? Why or why not?

   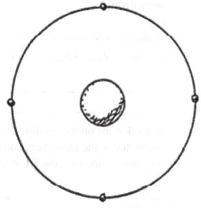

   Figure 2

   e. Are all four **V** vectors the same length? Why or why not?

   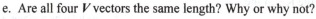

   f. What is the angle between your **F** and **V** vectors? _____

   g. Is there any component of **F** along **V**? _____

   h. What does this tell you about the work the force of gravity does on the satellite?

   _____

   i. Does the KE of the satellite in Figure 2 remain constant, or does it vary? _____

   j. Does the PE of the satellite remain constant, or does it vary? _____

⌐Conceptual Integrated Science Third Edition

Circular and Elliptical Orbits—continued

II. Elliptical Orbits

3. Figure 3 shows a satellite in elliptical orbit.

a. Repeat the procedure you used for the circular orbit, drawing vectors *F* and *V* for each position, including proper labeling. Show equal magnitudes with equal lengths, and greater magnitudes with greater lengths, but don't bother making the scale accurate.

b. Are your vectors *F* all the same magnitude?
Why or why not?

_____

_____

c. Are your vectors *V* all the same magnitude?
Why or why not?

_____

_____

_____

_____

d. Is the angle between vectors *F* and *V* everywhere the same, or does it vary?

_____

e. Are there places where there is a component of *F* along *V*?

_____

_____

f. Is work done on the satellite when there is a component of *F* along and in the same direction of *V*, and if so, does this increase or decrease the KE of the satellite?

_____

g. When there is a component of *F* along and opposite to the direction of *V*, does this increase or decrease the KE of the satellite?

_____

_____

h. What can you say about the sum KE + PE along the orbit?

_____

_____

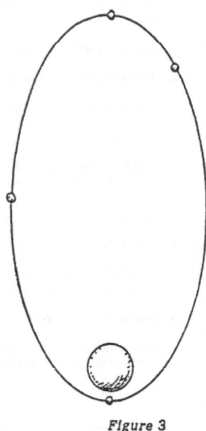

Figure 3

Be very, very careful when placing both velocity and force vectors on the same diagram. Not a good practice, for one may construct the resultant of the vectors—ouch!

Conceptual Integrated Science Third Edition

Chapter 5: Gravity
Mechanics Overview

1. The sketch shows the elliptical path described by a satellite about Earth. In which of the marked positions, A–D, (put S for "same everywhere") does the satellite experience the maximum

   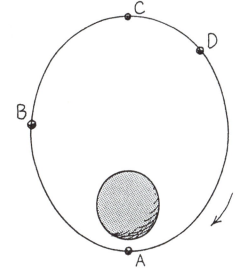

   a. gravitational force? _____

   b. speed? _____

   c. velocity? _____

   d. momentum? _____

   e. kinetic energy? _____

   f. gravitational potential energy? _____

   g. total energy (KE + PE)? _____

   h. acceleration? _____

2. Answer the above questions for a satellite in circular orbit.

   a. _____ b. _____ c. _____ d. _____ e. _____ f. _____ g. _____ h. _____

3. In which position(s) is there momentarily no work done on the satellite by the force of gravity? Why?

   _____

4. Work changes energy. Let the equation for work, $W = Fd$, guide your thinking on these questions. Defend your answers in terms of $W = Fd$.

   a. In which position will a several-minutes thrust of rocket engines do the most work on the satellite and give it the greatest change in kinetic energy?

   _____

   b. In which position will a several-minutes thrust of rocket engines do the most work on the *exhaust gases* and give the *exhaust gases* the greatest change in kinetic energy?

   _____

   c. In which position will a several-minutes thrust of rocket engines give the satellite the least boost in kinetic energy?

   _____

   _____

Conceptual Integrated Science — Third Edition

Chapter 6: Heat

Temperature Mix

1. You apply heat to 1 L of water and raise its temperature by 10°C. If you add the same quantity of heat to 2 L of water, how much will the temperature rise? To 3 L of water?

   Record your answers on the blanks in the drawing at the right. (Hint: Heat transferred is directly proportional to it temperature change, $Q = mc\Delta T$.)

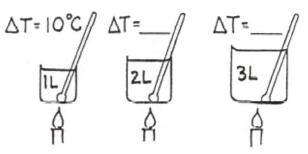

2. A large bucket contains 1 L of 20°C water.

   a. What will be the temperature of the mixture when 1 L of 20°C water is added?

   _____

   b. What will be the temperature of the mixture when 1 L of 40°C water is added?

   _____

   c. If 2 L of 40°C water were added, would the temperature of the mixture be greater or less than 30°C?

   _____

3. A red-hot iron kilogram mass is put into 1 L of cool water. Mark each of the following statements true (T) or false (F). (Ignore heat transfer to the container.)

   a. The increase in the water temperature is equal to the decrease in the iron's temperature.

   _____

   b. The quantity of heat gained by the water is equal to the quantity of heat lost by the iron.

   _____

   c. The iron and the water will both reach the same temperature.

   _____

   d. The final temperature of the iron and water is about halfway between the initial temperatures of each.

   _____

4. *True or False:* When Queen Elizabeth throws the last sip of her tea over Queen Mary's rail, the ocean gets a little warmer. _____

Conceptual Integrated Science
Third Edition

Chapter 6: Heat

Absolute Zero

A mass of air is contained so that the volume can change but the pressure remains constant. Table 1 shows air volumes at various temperatures when the air is heated slowly.

1. Plot the data in Table 1 on the graph and connect the points.

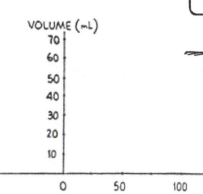

Table 1

TEMP. (°C)	VOLUME (mL)
0	50
25	55
50	60
75	65
100	70

2. The graph shows how the volume of air varies with temperature at constant pressure. The straightness of the line means that the air expands uniformly with temperature. From your graph, you can predict what will happen to the volume of air when it is cooled.

Extrapolate (extend) the straight line of your graph to find the temperature at which the volume of the air would become zero. Mark this point on your graph. Estimate this temperature: _____

3. Although air would liquify before cooling to this temperature, the procedure suggests that there is a lower limit to how cold something can be. This is the absolute zero of temperature.

Careful experiments show that absolute zero is _____ °C.

4. Scientists measure temperature in *kelvins* instead of degrees Celsius, where the absolute zero of temperature is 0 kelvins. If you relabeled the temperature axis on the graph in Question 1 so that it shows temperature in kelvins, would your graph look like the one below? _____

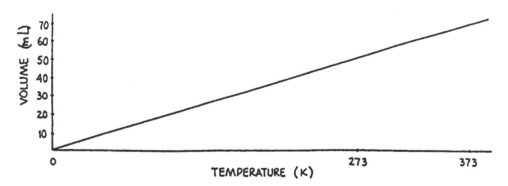

Conceptual Integrated Science — Third Edition

Chapter 6: Heat
Thermal Expansion

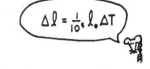
$$\Delta l = \tfrac{1}{10^5} \, l_\circ \Delta T$$

1. Steel expands by about 1 part in 100,000 for each 1°C increase in temperature.

   a. How much longer will a piece of steel 1000 mm long (1 meter) be when its temperature is increased by 10°C? _____

   b. How much longer will a piece of steel 1000 m long (1 kilometer) be when its temperature is increased by 10°C? _____

   c. You place yourself between a wall and the end of a 1-m steel rod when the opposite end is securely fastened as shown. No harm comes to you if the temperature of the rod is increased a few degrees. Discuss the consequences of doing this with a rod many meters long.

   _____

   _____

2. The Eiffel Tower in Paris is 298 meters high. On a cold winter night, it is shorter than on a hot summer day. What is its change in height for a 30°C temperature difference?

   _____

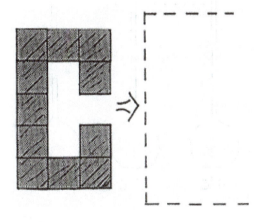

3. Consider a gap in a piece of metal. Does the gap become wider or narrower when the metal is heated? (Consider the piece of metal made up of 11 blocks—if the blocks are individually heated, each is slightly larger. Make a sketch of them, slightly enlarged, beside the sketch shown.)

   _____

4. The equatorial radius of Earth is about 6370 km. Consider a 40,000-km long steel pipe that forms a giant ring that fits snugly around Earth's equator. Suppose people all along its length breathe on it so as to raise its temperature by 1°C. The pipe gets longer. It is also no longer snug. How high does it stand above the ground? (Hint: Concentrate on the radial distance.)

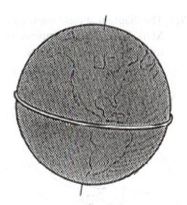

Conceptual Integrated Science
Third Edition

Thermal Expansion—continued

5. A weight hangs above the floor from the copper wire. When a candle is moved along the wire and heats it, what happens to the height of the weight above the floor? Why?

_____

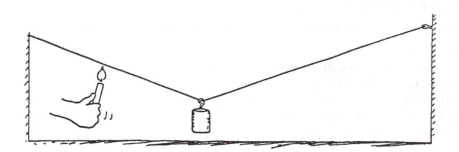

6. The levels of water at 0°C and 1°C are shown below in the first two flasks. At these temperatures there is microscopic slush in the water. There is slightly more slush at 0°C than at 1°C. As the water is heated, some of the slush collapses as it melts, and the level of the water falls in the tube. That's why the level of water is slightly lower in the 1°C tube. Make rough estimates and sketch in the appropriate levels of water at the other temperatures shown. What is important about the level when the water reaches 4°C?

_____

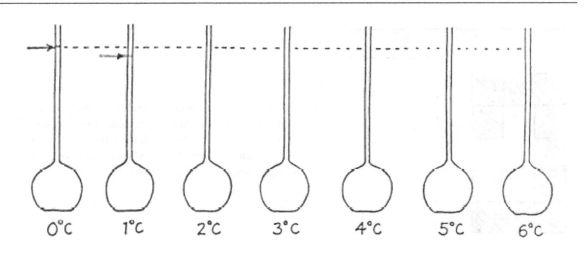

7. The diagram at right shows an ice-covered pond. Mark the probable temperatures of water at the top and bottom of the pond.

I CAN'T GET THIS METAL LID OFF THE JAR ⋯ SHOULD I HEAT THE LID OR COOL IT? WHY? _____

WHICH WILL WEIGH MORE, 1 LITER OF ICE OR 1 LITER OF WATER? _____

Name _____ Date _____

Conceptual Integrated Science — Third Edition

Chapter 6: Heat

Transmission of Heat

1. The tips of both brass rods are held in the gas flame. *Mark the following true (T) or false (F).*

   a. Heat is conducted only along Rod A. _____

   b. Heat is conducted only along Rod B. _____

   c. Heat is conducted equally along both Rod A and Rod B.

   _____

   d. The idea that "heat rises" applies to heat transfer by *convection*, not by *conduction*. _____

2. Why does a bird fluff its feathers to keep warm on a cold day?

   _____

3. Why does a down-filled sleeping bag keep you warm on a cold night? Why is it useless if the down is wet?

   _____

4. What does *convection* have to do with the holes in the shade of the desk lamp?

   _____

5. When hot water rapidly evaporates, the result can be dramatic. Consider 4 g of boiling water spread over a large surface so that 1 g rapidly evaporates. Suppose further that the surface and surroundings are very cold so that all 540 calories for evaporation come from the remaining 3 g of water.

   a. How many calories are taken from each gram of water?

   _____

   b. How many calories are released when 1 g of 100°C water cools to 0°C?

   _____

   c. How many calories are released when 1 g of 0°C water changes to 0°C ice?

   _____

   d. What happens in this case to the remaining 3 g of boiling water when 1 g rapidly evaporates?

   _____

   _____

Conceptual Integrated Science
Third Edition

Chapter 7: Electricity and Magnetism
Electric Potential

Just as PE transforms into KE for a mass lifted against the gravitation field (left), the electric PE of an electric charge transforms into other forms of energy when it changes location in an electric field (right). In both cases, how does the KE acquired compare with the decrease in PE?

_____

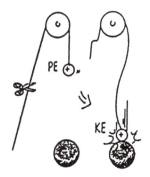

Complete the following statements:

A force compresses the spring. The work done in compression is the product of the average force and the distance moved: $W = Fd$. This work increases the PE of the spring.

Similarly, a force pushes the charge (call it a *test charge*) closer to the charged sphere. The work done in moving the test charge is the product of the average _____ and the _____ moved: W = _____. This work _____ the PE of the test charge.

If the test opcharge is released, it will be repelled and fly past the starting point. Its gain in KE at this point is _____ to its decrease in PE.

At any point, a greater amount of test charge means a greater amount of PE, but not a greater amount of PE *per amount* of charge. The quantities PE (measured in joules) and $\frac{PE}{charge}$ (measured in volts) are

different concepts.

By definition: Electric Potential $= \frac{PE}{charge}$. 1 volt $= \frac{1\ joule}{1\ coulomb}$. So, 1 C of charge with a PE of 1 J

has an electric potential of 1 V; 2 C of charge with a PE of 2 J has an electric potential of _____ V.

If a conductor connected to the terminal of a battery has an electric potential of 12 V, then each coulomb of charge on the conductor has a PE of _____ J.

You do very little work in rubbing a balloon on your hair to charge it. The PE of several thousand billion electrons (about one-millionth coulomb [10^{-6}C]) transferred may be a thousandth of a joule [10^{-3}J]. Impressively, however, the electric potential of the balloon is about _____ V!

Why is contact with a balloon charged to thousands of volts not as dangerous as contact with household 110 V?

_____

_____

Conceptual Integrated Science
Third Edition

Chapter 7: Electricity and Magnetism
Series Circuits

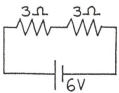

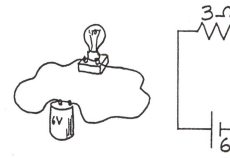

1. The simple circuit is a 6-V battery that pushes charge through a single lamp that has a resistance of 3 Ω. According to Ohm's law, the current in the lamp (and therefore the whole circuit) is _____ A.

2. If a second identical lamp is added, the 6-V battery must push charge through a total resistance of _____ Ω. The current in the circuit is then _____ A.

3. If a third identical lamp is added in series, the total resistance of the circuit (neglecting any internal resistance in the battery) is _____ Ω.

4. The current through all three lamps in series is _____ A. The current through each individual lamp is _____ A.

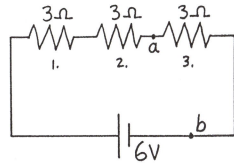

5. Does current in the lamps occur simultaneously, or does charge flow first through one lamp, then the other, and finally the last, in turn? _____

6. Does current flow *through* a resistor, or *across* a resistor? _____ Is voltage established *through* a resistor, or *across* a resistor? _____

7. The voltage across all three lamps in the series is 6-V. The voltage (or commonly, *voltage drop*) across each individual lamp is ____2____ V.

8. Suppose a wire connects points *a* and *b* in the circuit. The voltage drop across lamp 1 is now _____ V, across lamp 2 is _____ V, and across lamp 3 is _____ V. So, the current through lamp 1 is now _____ A, through lamp 2 is _____ A, and through lamp 3 is _____ A. The current in the battery (neglecting internal battery resistance) is _____ A.

9. Which circuit dissipates more power: the 3-lamp circuit or the 2-lamp circuit? (Another way of asking this is, which circuit would glow brightest and be best seen on a dark night from a great distance?) Defend your answer.

_____
_____
_____

─Conceptual Integrated Science─ **Third Edition**

Chapter 7: Electricity and Magnetism
Parallel Circuits

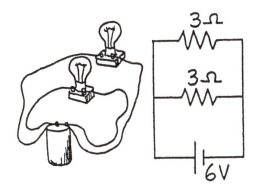

3 Ω

3 Ω

6V

1. In the circuit shown to the left there is a voltage drop of 6V across each 3-Ω lamp. By Ohm's law, the current in each lamp is _____ A. The current through the battery is the sum of the currents in the lamps: _____ A.

> THE SUM OF THE CURRENTS IN THE TWO BRANCH PATHS EQUALS THE CURRENT BOTH BEFORE AND AFTER IT DIVIDES!
> WATER FLOW

2. Fill in the current in the eight blank spaces in the view of the same circuit shown again on the right.

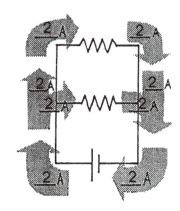

3. Suppose a third identical lamp is added in parallel to the circuit. Sketch a schematic diagram of the 3-lamp circuit in the space on the right.

4. For the three identical lamps in parallel, the voltage drop across each lamp is _____V. The current through each lamp is _____A. The current through the battery is now _____A. Is the circuit resistance now greater or lesser than before the third lamp was added? Explain.

_____

5. Which circuit dissipates more power: the 3-lamp circuit or the 2-lamp circuit? (Another way of asking this is, which circuit would glow brightest and be best seen on a dark night from a great distance?) Defend your answer and compare this to the similar case for 2- and 3-lamp series circuits.

_____

_____

Name _____   Date _____

Conceptual Integrated Science
Third Edition

Chapter 7: Electricity and Magnetism
Compound Circuits

The table beside circuit *a* below shows the current through each resistor, the voltage across each resistor, and the power dissipated as heat in each resistor. Find the similar correct values for circuits *b, c,* and *d,* and put your answers in the tables shown.

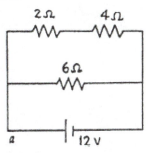

RESISTANCE	CURRENT ×	VOLTAGE =	POWER
2 Ω	2 A	4 V	8 W
4 Ω	2 A	8 V	16 W
6 Ω	2 A	12 V	24 W

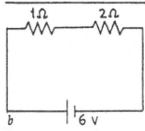

RESISTANCE	CURRENT ×	VOLTAGE =	POWER
1 Ω			
2 Ω			

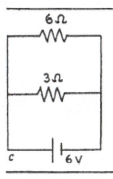

RESISTANCE	CURRENT ×	VOLTAGE =	POWER
6 Ω			
3 Ω			

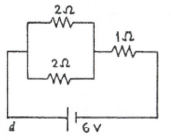

RESISTANCE	CURRENT ×	VOLTAGE =	POWER
2 Ω			
2 Ω			
1 Ω			

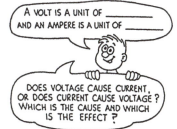

A VOLT IS A UNIT OF _____
AND AN AMPERE IS A UNIT OF _____

DOES VOLTAGE CAUSE CURRENT,
OR DOES CURRENT CAUSE VOLTAGE ?
WHICH IS THE CAUSE AND WHICH
IS THE EFFECT ?

Name _____   Date _____

Conceptual Integrated Science — Third Edition

Chapter 7: Electricity and Magnetism
Magnetism

Fill in each blank with the appropriate word:

1. Attraction or repulsion of charges depends on their *signs*: positives or negatives. Attraction or repulsion of magnets depends on their magnetic _____: _____ or _____.

2. Opposite poles attract; like poles _____.

3. A magnetic field is produced by the _____ of electric charge.

4. Clusters of magnetically aligned atoms are magnetic _____.

5. A magnetic _____ surrounds a current-carrying wire.

6. When a current-carrying wire is made to form a coil around a piece of iron, the result is an

   _____

7. A charged particle moving in a magnetic field experiences a deflecting _____ that is maximum when the charge moves _____ to the field.

8. A current-carrying wire experiences a deflecting _____ that is maximum when the wire and magnetic field are _____ to one another.

9. A simple instrument designed to detect electric current is the _____; when calibrated to measure current, it is an _____; when calibrated to measure voltage, it is a _____

10. The largest size magnet in the world is the _____ itself.

Conceptual Integrated Science — Third Edition

Chapter 7: Electricity and Magnetism
Field Patterns

1. The illustration below is similar to Figure 7.32 in your textbook. Iron filings trace out patterns of magnetic field lines about a bar magnet. In the field are some magnetic compasses. The compass needle in only one compass is shown. Draw in the needles with proper orientation in the other compasses.

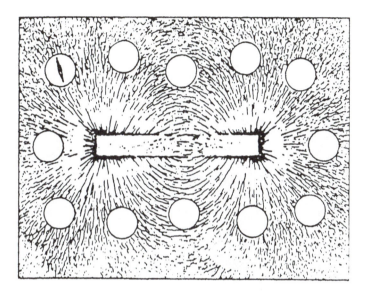

2. The illustration below is similar to Figure 7.37b in your textbook. Iron filings trace out the magnetic field pattern about the loop of current-carrying wire. Draw in the compass needle orientations for all the compasses.

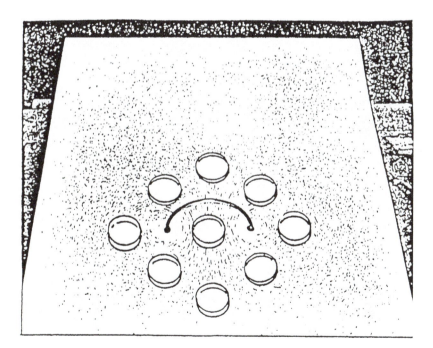

Conceptual Integrated Science — Third Edition

Chapter 7: Electricity and Magnetism
Electromagnetism

1. Early investigators discovered that magnetism and electricity are

    (related)   (independent of each other).

   Magnetism is produced by

    (batteries)   (the motion of electric charges).

Faraday and Henry discovered that electric current can be produced by

    (batteries)   (motion of a magnet).

More specifically, voltage is induced in a loop of wire if there is a change in the

    (batteries)   (magnetic field in the loop).

This phenomenon is called

    (electromagnetism)   (electromagnetic induction).

2. When a magnet is plunged in and out of a coil of wire, voltage is induced in the coil. If the rate of the in-and-out motion of the magnet is doubled, the induced voltage

    (doubles)   (halves)   (remains the same).

   If instead, the number of loops in the coil is doubled, the induced voltage

    (doubles)   (halves)   (remains the same).

3. A rapidly changing magnetic field in any region of space induces a rapidly changing

    (electric field)   (magnetic field)   (gravitational field),

   which in turn induces a rapidly changing

    (magnetic field)   (electric field)   (baseball field).

   This generation and regeneration of electric and magnetic fields makes up

    (electromagnetic waves)   (sound waves)   (both of these).

Name _____   Date _____

Conceptual Integrated Science — Third Edition

Chapter 8: Waves—Sound and Light
Vibration and Wave Fundamentals

1. A sine curve that represents a transverse wave is drawn below. With a ruler, measure the wavelength and amplitude of the wave.

      a. Wavelength = _____   b. Amplitude = _____

2. A girl on a playground swing makes a complete to-and-fro swing each 2 seconds. The frequency of swing is

    (0.5 hertz)   (1 hertz)   (2 hertz)

and the period is

    (0.5 second)   (1 second)   (2 seconds).

3. *Complete the following statements:*

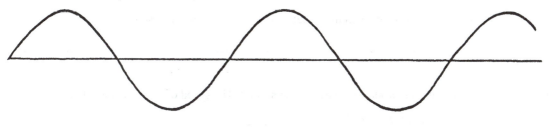

THE PERIOD OF A 440-HERTZ SOUND WAVE IS _____ SECOND(S).

A MARINE WEATHER STATION REPORTS WAVES ALONG THE SHORE THAT ARE 8 SECONDS APART. THE FREQUENCY OF THE WAVES IS THEREFORE _____ HERTZ.

4. The annoying sound from a mosquito occurs because it beats its wings at the average rate of 600 wingbeats per second.

    a. What is the frequency of the soundwaves?

    _____

    b. What is the wavelength? (Assume the speed of sound is 340 m/s.)

    _____

Conceptual Integrated Science
Third Edition

Vibration and Wave Fundamentals—continued

5. A machine gun fires 10 rounds per second. The speed of the bullets is 300 m/s.

   a. What is the distance in the air between the flying bullets? _____

   b. What happens to the distance between the bullets if the rate of fire is increased?

   _____

6. Consider a wave generator that produces 10 pulses per second. The speed of the waves is 300 cm/s.

   a. What is the wavelength of the waves? _____

   b. What happens to the wavelength if the frequency of pulses is increased?

   _____

7. The bird at the right watches the waves. If the portion of a wave between 2 crests passes the pole each second, what is the speed of the wave?

   _____

   What is its period?

   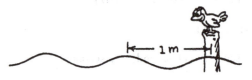

   _____

8. If the distance between crests in the above question were 1.5 meters, and 2 crests pass the pole each second, what would be the speed of the wave?

   _____

   What would be its period?

   _____

9. When an automobile moves toward a listener, the sound of its horn seems relatively

   (low pitched)   (normal)   (high pitched).

   When moving away from the listener, its horn seems

   (low pitched)   (normal)   (high pitched).

10. The changed pitch of the Doppler effect is due to changes in

   (wave speed)   (wave frequency).

─Conceptual Integrated Science─ Third Edition

Chapter 8: Waves—Sound and Light
Color

The sketch to the right shows the shadow of an instructor in front of a white screen in a dark room. The light source is red, so the screen looks red and the shadow looks black. Color the sketch, or label the colors with a pen or pencil.

A green lamp is added and makes a second shadow. The shadow cast by the red lamp is no longer black, but is illuminated by green light, so it is green. Color or mark it green. The shadow cast by the green lamp is not black because it is illuminated by the red lamp. Indicate its color. Do the same for the background, which receives a mixture of red and green light.

A blue lamp is added and three shadows appear. Indicate the appropriate colors of the shadows and the background.

The lamps are placed a bit closer together so the shadows overlap. Indicate the colors of all screen areas.

Color—continued

If you have colored pencils or markers, have a go at these.

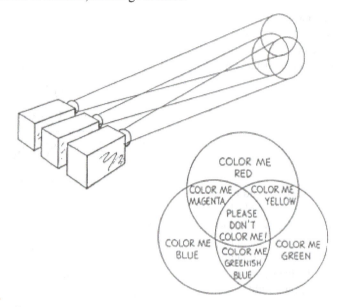

Conceptual Integrated Science Third Edition

Chapter 8: Waves—Sound and Light

Diffraction and Interference

Shown below are concentric solid and dashed circles, each different in radius by 1 cm. Consider the circular pattern of a top view of water waves, where the solid circles are crests and the dashed circles are troughs.

1. Draw another set of the same concentric circles with a compass. Choose any part of the paper for your center (except the present central point). Let the circles run off the edge of the paper.

2. Find where a dashed line crosses a solid line and draw a large dot at the intersection. Do this for ALL places where a solid and dashed line intersect.

3. With a wide felt marker, connect the dots with smooth lines. These *nodal lines* lie in regions where the waves have cancelled—where the crest of one wave overlaps the trough of another.

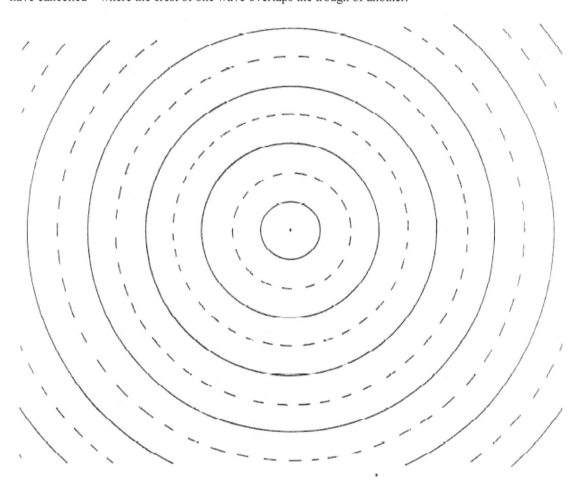

Conceptual Integrated Science
Third Edition

Chapter 8: Waves—Sound and Light
Reflection

1. Light from a flashlight shines on a mirror and illuminates one of the cards. Draw the reflected beam to indicate the illuminated card.

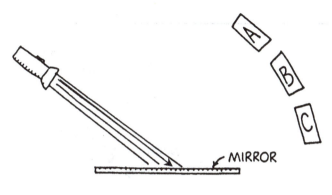

2. A periscope has a pair of mirrors in it. Draw the light path from the object "O" to the eye of the observer.

3. The ray diagram below shows the extension of one of the reflected rays from the plane mirror. Complete the diagram by (1) carefully drawing the three other reflected rays and (2) extending them behind the mirror to locate the image of the flame. (Assume the candle and image are viewed by an observer on the left.)

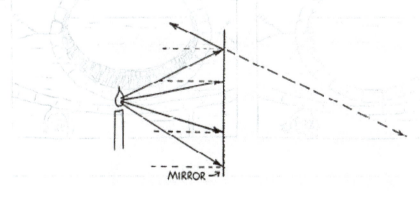

Conceptual Integrated Science
Third Edition

Reflection—continued

4. The ray diagram below shows the reflection of one of the rays that strikes the parabolic mirror. Notice that the law of reflection is observed, and the angle of incidence (from the normal, the dashed line) equals the angle of reflection (from the normal). Complete the diagram by drawing the reflected rays of the other three rays that are shown. (Do you see why parabolic mirrors are used in automobile headlights?)

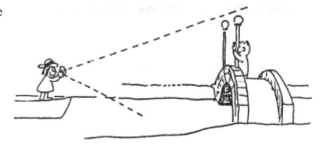

MIRROR

5. A girl takes a photograph of the bridge as shown. Which of the two sketches below correctly shows the reflected view of the bridge? Defend your answer.

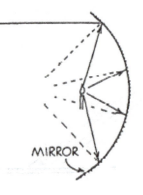

_____

_____

_____

_____

Conceptual Integrated Science Third Edition

Chapter 8: Waves—Sound and Light

Refraction—Part 1

1. A pair of toy cart wheels are rolled obliquely from a smooth surface onto two plots of grass—a rectangular plot as shown at the left, and a triangular plot as shown the right. The ground is on a slight incline, so that after slowing down in the grass, the wheels speed up again when emerging on the smooth surface. Finish each sketch and show some positions of the wheels inside the plots and on the other side. Clearly indicate their paths and directions of travel.

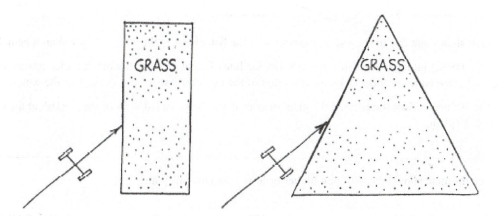

2. Red, green, and blue rays of light are incident upon a glass prism as shown. The average speed of red light in the glass is less than in air, so the red ray is refracted. When it emerges into the air it regains its original speed and travels in the direction shown. Green light takes longer to get through the glass. Because of its slower speed, it is refracted as shown. Blue light travels even slower in glass. Complete the diagram by estimating the path of the blue ray.

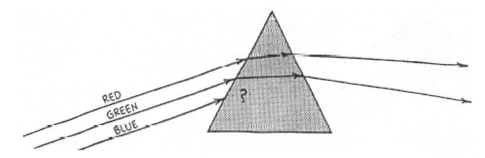

3. Below, we consider a prism-shaped hole in a piece of glass—that is, an "air prism." Complete the diagram showing likely paths of the beams of red, green, and blue light as they pass through this "prism" and back to glass.

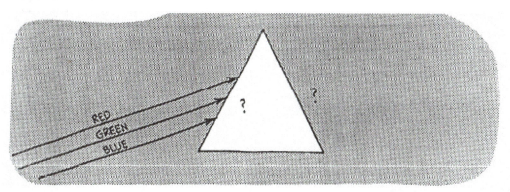

Conceptual Integrated Science — Third Edition

Refraction—Part 1—continued

4. Light of different colors diverges when emerging from a prism. Newton showed that with a second prism he could make the diverging beams become parallel again. Which placement of the second prism will do this?

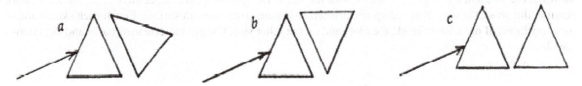

5. The sketch shows that due to refraction, the man sees the fish closer to the water surface than it actually is.

   a. Draw a ray beginning at the fish's eye to show the line of sight of the fish when it looks upward at 50° to the normal at the water surface. Draw the direction of the ray after it meets the surface of the water.

   b. At the 50° angle, does the fish see the man, or does it see the reflected view of the starfish at the bottom of the pond? Explain.

   _____

   c. To see the man, should the fish look higher or lower than the 50° path?

   _____

   d. If the fish's eye were barely above the water surface, it would see the world above in a 180° view, horizon to horizon. The fish-eye view of the world above as seen beneath the water, however, is very different. Due to the 48° critical angle of water, the fish sees a normally 180° horizon-to-horizon view compressed within an angle of _____.

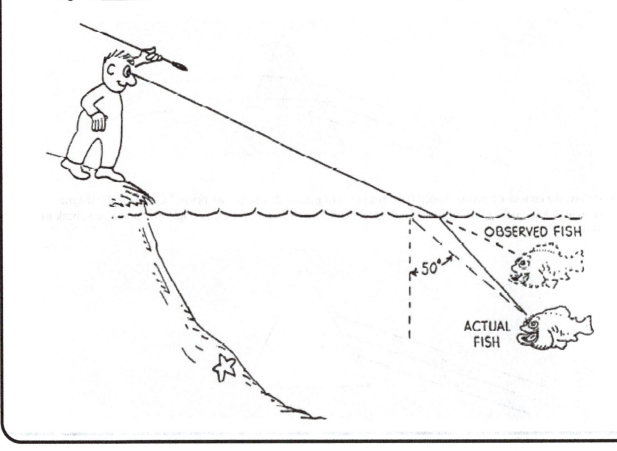

Conceptual Integrated Science
Third Edition

Chapter 8: Waves—Sound and Light
Refraction—Part 2

1. The sketch to the right shows a light ray moving from air into water, at 45° to the normal. Which of the three rays indicated with capital letters is most likely the light ray that continues inside the water?

   _____

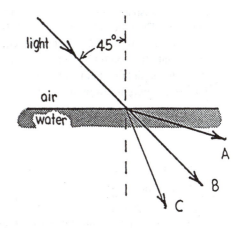

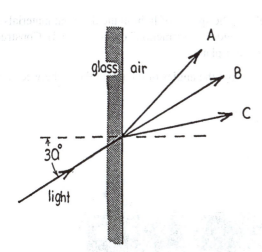

2. The sketch on the left shows a light ray moving from glass into air, at 30° to the normal. Which of the three is most likely the light ray that continues in the air?

   _____

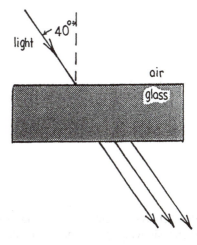

3. To the right, a light ray is shown moving from air into a glass block, at 40° to the normal. Which of the three rays is most likely the light ray that travels in the air after emerging from the opposite side of the block?

   _____

   Sketch the path the light would take inside the glass.

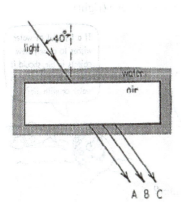

4. To the left, a light ray is shown moving from water into a rectangular block of air (inside a thin-walled plastic box), at 40° to the normal. Which of the three rays is most likely the light ray that continues into the water on the opposite side of the block?

   _____

   Sketch the path the light would take inside the air.

Conceptual Integrated Science ── Third Edition

Refraction—Part 2—continued

5. The two transparent blocks (right) are made of different materials. The speed of light in the left block is greater than the speed of light in the right block. Draw an appropriate light path through and beyond the right block. Is the light that emerges displaced more or less than light emerging from the left block?

_____

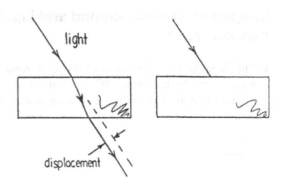

6. Light from the air passes through plates of glass and plastic below. The speeds of light in the different materials is shown to the right (these different speeds are often implied by the "index of refraction" of the material). Construct a rough sketch showing an appropriate path through the system of four plates.

Compared with the 50° incident ray at the top, what can you say about the angles of the ray in the air between and below the block pairs?

_____

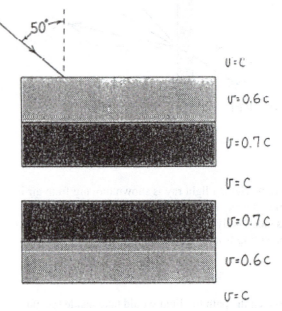

7. Parallel rays of light are refracted as they change speed in passing from air into the eye (left). Construct a rough sketch showing appropriate light paths when parallel light under water meets the same eye (right).

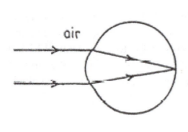

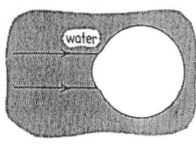

If a fish out of water wishes to clearly view objects in air, should it wear goggles filled with water or with air?

8. Why do we need to wear a face mask or goggles to see clearly when under water?

_____

Conceptual Integrated Science — Third Edition

Chapter 8: Waves—Sound and Light
Wave–Particle Duality

1. To say that light is quantized means that light is made up of

   (elemental units) (waves).

2. Compared with photons of low-frequency light, photons of higher-frequency light have more

   (energy) (speed) (quanta).

3. The photoelectric effect supports the

   (wave model of light) (particle model of light).

4. The photoelectric effect is evident when light shone on certain photosensitive materials ejects

   (photons) (electrons).

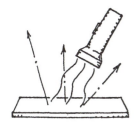

5. The photoelectric effect is more effective with violet light than with red light because the photons of violet light

   (resonate with the atoms in the material)

   (deliver more energy to the material)

   (are more numerous).

6. According to the wave model of matter, a beam of light and a beam of electrons

   (are fundamentally different) (are similar).

7. According to De Broglie, the greater the speed of an electron beam, the

   (greater is its wavelength) (shorter is its wavelength).

8. The discreteness of the energy levels of electrons about the atomic nucleus is best understood by considering the electron to be a

   (wave) (particle).

9. Heavier atoms are not appreciably larger in size than lighter atoms. The main reason for the similarity of sizes is the greater nuclear charge

   (pulls surrounding electrons into tighter orbits)

   (holds more electrons about the atomic nucleus)

   (produces a denser atomic structure).

A QUANTUM MECHANIC!

10. Whereas in the everyday macroworld the study of motion is called *mechanics,* in the microworld the study of quanta is called

   (Newton mechanics) (quantum mechanics).

─Conceptual Integrated Science ──── Third Edition

Chapter 9: Atoms and the Periodic Table

Subatomic Particles

Three fundamental particles of the atom are the _____, _____, and _____. At the center of each atom lies the atomic _____, which consists of _____ and _____. The **atomic number** refers to the number of _____ in the nucleus. All atoms of the same element have the same number of _____, hence, the same atomic number.

Isotopes are atoms that have the same number of _____, but a different number of _____. An isotope is identified by its **atomic mass number,** which is the total number of _____ and _____ in the nucleus. A carbon isotope that has 6 _____ and 6 _____ is identified as carbon-12, where 12 is the atomic mass number. A carbon isotope having 6 _____ and 8 _____, on the other hand, is carbon-14.

1. Complete the following table:

Isotope	Number of...		
	Electrons	Protons	Neutrons
Hydrogen-1	1		
Chlorine-36		17	
Nitrogen-14			7
Potassium-40	19		
Arsenic-75		33	
Gold-197			118

2. Which results in a more valuable product—*adding* or *subtracting* protons from gold nuclei?

3. Which has more mass, a helium atom or a neon atom?

4. Which has a greater number of atoms, a gram of helium or a gram of neon?

Name _____ Date _____

Conceptual Integrated Science
Third Edition

Chapter 10: The Atomic Nucleus and Radioactivity
Radioactivity

1. Complete the following statements.

   a. A lone neutron spontaneously decays into a proton plus an _____.

   b. Alpha and beta rays are made of streams of particles, whereas gamma rays are streams of
      _____.

   c. An electrically charged atom is called an _____.

   d. Different _____ of an element are chemically identical but differ in the number of
      neutrons in the nucleus.

   e. Transuranic elements are those beyond atomic number _____.

   f. If the amount of a certain radioactive sample decreases by half in four weeks, in four more weeks the amount
      remaining should be _____ the original amount.

   g. Water from a natural hot spring is warmed by _____ inside Earth.

2. The gas in the little girl's balloon is made up of former alpha and beta particles produced by radioactive decay.

   a. If the mixture is electrically neutral, how many more beta particles
      than alpha particles are in the balloon?

      _____

      _____

   b. Why is your answer not "same"?

      _____

      _____

      _____

   c. Why are the alpha and beta particles no longer harmful to the child?

      _____

      _____

   d. What element does this mixture make?

      _____

Conceptual Integrated Science
Third Edition

Radioactivity—continued

Draw in a decay-scheme diagram below, similar to Figure 10.17 in your text. In this case, you begin at the upper right with U-235 and end up with a different isotope of lead. Use the table at the left and identify each element in the series by its chemical symbol.

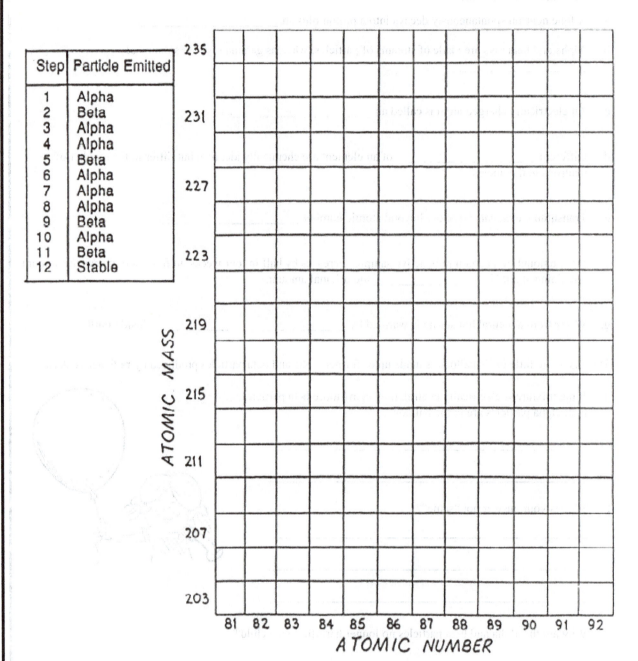

Step	Particle Emitted
1	Alpha
2	Beta
3	Alpha
4	Alpha
5	Beta
6	Alpha
7	Alpha
8	Alpha
9	Beta
10	Alpha
11	Beta
12	Stable

ATOMIC MASS

ATOMIC NUMBER

Which isotope is the final product? _____

Conceptual Integrated Science
Third Edition

Chapter 10: The Atomic Nucleus and Radioactivity
Radioactive Half-Life

You and your classmates will now play the "half-life game." Each of you should have a coin to shake inside cupped hands. After it has been shaken for a few seconds, the coin is tossed on the table or on the floor. Students with tails up fall out of the game. Only those who consistently show heads remain in the game. Finally, everybody has tossed a tail and the game is over.

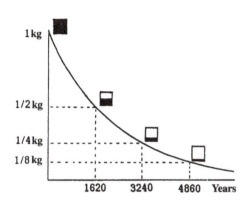

1. The graph to the left shows the decay of Radium-226 with time. Note that each 1620 years, half remains (the rest changes to other elements). In the grid below, plot the number of students left in the game after each toss. Draw a smooth curve that passes close to the points on your plot. What is the similarity of your curve with that of the curve of Radium-226?

    _____

    _____

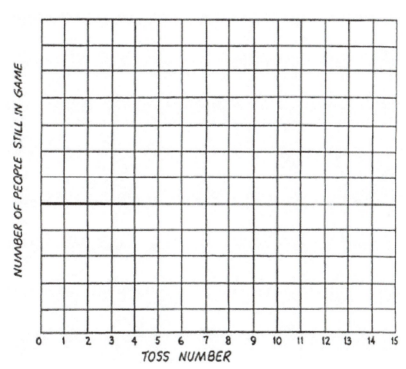

2. Was the person to last longest in the game *lucky,* with some sort of special powers to guide the long survival? What test could you make to decide the answer to this question?

_____

Chapter 10: The Atomic Nucleus and Radioactivity
Nuclear Fission and Fusion

1. Complete the table for a chain reaction in which two neutrons from each step individually cause a new reaction.

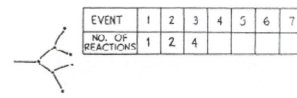

EVENT	1	2	3	4	5	6	7
NO. OF REACTIONS	1	2	4				

2. Complete the table for a chain reaction in which three neutrons from each reaction cause a new reaction.

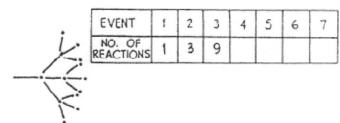

EVENT	1	2	3	4	5	6	7
NO. OF REACTIONS	1	3	9				

3. Complete these beta reactions, which occur in a fission breeder reactor.

$$^{239}_{92}U \longrightarrow \underline{\hspace{1.5cm}} + ^{0}_{-1}e$$

$$^{239}_{93}Np \longrightarrow \underline{\hspace{1.5cm}} + ^{0}_{-1}e$$

4. Complete the following fission reactions.

$$^{1}_{0}n + ^{235}_{92}U \longrightarrow ^{143}_{54}Xe + ^{90}_{38}Sr + \underline{\hspace{1cm}} \left(^{1}_{0}n\right)$$

$$^{1}_{0}n + ^{235}_{92}U \longrightarrow ^{152}_{60}Nd + \underline{\hspace{1.5cm}} + 4\left(^{1}_{0}n\right)$$

$$^{1}_{0}n + ^{239}_{94}Pu \longrightarrow \underline{\hspace{1.5cm}} + ^{97}_{40}Zr + 2\left(^{1}_{0}n\right)$$

5. Complete the following fusion reactions.

$$^{2}_{1}H + ^{2}_{1}H \longrightarrow ^{3}_{2}He + \underline{\hspace{1cm}}$$

$$^{2}_{1}H + ^{3}_{1}H \longrightarrow ^{4}_{2}He + \underline{\hspace{1cm}}$$

Name _____ Date _____

Conceptual Integrated Science — Third Edition

Chapter 10: The Atomic Nucleus and Radioactivity

Nuclear Reactions

Complete these nuclear reactions:

1. $^{230}_{90}\text{Th} \longrightarrow \, ^{226}_{88}\text{Ra} + \underline{}$

2. $^{218}_{85}\text{At} \longrightarrow \underline{} + \, ^{4}_{2}\text{He}$

3. $^{14}_{6}\text{C} \longrightarrow \underline{} + \, ^{14}_{7}\text{N}$

4. $^{80}_{35}\text{Br} \longrightarrow \, ^{80}_{36}\text{Kr} + \underline{}$

5. $^{214}_{83}\text{Bi} \longrightarrow \, ^{4}_{2}\text{He} + \underline{}$

6. $^{212}_{83}\text{Bi} \longrightarrow \, ^{0}_{-1}\text{e} + \underline{}$

7. $^{80}_{35}\text{Br} \longrightarrow \, ^{0}_{-1}\text{e} + \underline{}$

8. $^{80}_{35}\text{Br} \longrightarrow \, ^{0}_{+1}\text{e} + \underline{}$

9. $^{1}_{1}\text{H} + \, ^{7}_{3}\text{Li} \longrightarrow \, ^{4}_{2}\text{He} + \underline{}$

10. $^{2}_{1}\text{H} + \, ^{3}_{1}\text{H} \longrightarrow \, ^{4}_{2}\text{He} + \underline{}$

NUCLEAR PHYSICS --- IT'S THE SAME TO ME WITH THE FIRST TWO LETTERS INTERCHANGED!

Chapter 11: Investigating Matter

Melting Points of the Elements

There is a remarkable degree of organization in the periodic table. As discussed in your textbook, elements within the same atomic group (vertical column) share similar properties. Also, the chemical reactivity of an element can be deduced from its position in the periodic table. Two additional examples of the periodic table's organization are the melting points and densities of the elements.

The periodic table below shows the melting points of nearly all the elements. Note the melting points are not randomly oriented, but, with only a few exceptions, either gradually increase or decrease as you move in any particular direction. This can be clearly illustrated by color coding each element according to its melting point.

Use colored pencils to color in each element according to its melting point. Use the suggested color legend. Color lightly so that symbols and numbers are still visible.

Color	Temperature Range, °C	Color	Temperature Range, °C
Violet	-273 — -50	Yellow	1400 — 1900
Blue	-50 — 300	Orange	1900 — 2900
Cyan	300 — 700	Red	2900 — 3500
Green	700 — 1400		

Melting Points of the Elements (°C)

1	2	3	4	5	6	7	8	9	10	11	12	13	14	15	16	17	18
H -259																	He -272
Li 180	Be 1278											B 2079	C 3550	N -210	O -218	F -219	Ne -248
Na 97	Mg 648											Al 660	Si 1410	P 44	S 113	Cl -100	Ar -189
K 63	Ca 839	Sc 1541	Ti 1660	V 1890	Cr 1857	Mn 1244	Fe 1535	Co 1495	Ni 1453	Cu 1083	Zn 419	Ga 30	Ge 937	As 817	Se 217	Br -7	Kr -156
Rb 39	Sr 769	Y 1522	Zr 1852	Nb 2468	Mo 2617	Tc 2172	Ru 2310	Rh 1966	Pd 1554	Ag 961	Cd 320	In 156	Sn 231	Sb 630	Te 449	I 113	Xe -111
Cs 28	Ba 725	La 921	Hf 2227	Ta 2996	W 3410	Re 3180	Os 3045	Ir 2410	Pt 1772	Au 1064	Hg -38	Tl 303	Pb 327	Bi 271	Po 254	At 302	Rn -71
Fr 27	Ra 700	Ac 1050	--	--	--	--	--	--									

Lanthanides:

Ce 799	Pr 931	Nd 1021	Pm 1168	Sm 1077	Eu 822	Gd 1313	Tb 1356	Dy 1412	Ho 1474	Er 1159	Tm 1545	Yb 819	Lu 1663

Actinides:

Th 1750	Pa 1600	U 1132	Np 640	Pu 641	Am 994	Cm 1340	Bk --	Cf --	Es --	Fm --	Md --	No --	Lr --

1. Which elements have the highest melting points?

_____

2. Which elements have the lowest melting points?

_____

3. Which atomic groups tend to go from higher to lower melting points reading from top to bottom? (Identify each group by its group number.)

_____

4. Which atomic groups tend to go from lower to higher melting points reading from top to bottom?

_____

Conceptual Integrated Science — Third Edition

Chapter 11: Investigating Matter

Densities of the Elements

The periodic table below shows the densities of nearly all the elements. As with the melting points, the densities of the elements either gradually increase or decrease as you move in any particular direction. Use colored pencils to color in each element according to its density. Shown below is a suggested color legend. Color lightly so that symbols and numbers are still visible. (Note: All gaseous elements are marked with an asterisk and should be the same color. Their densities, which are given in units of g/L, are much less than the densities of nongaseous elements, which are given in units of g/mL.)

Color	Density (g/mL)	Color	Density (g/mL)
Violet	gaseous elements	Yellow	16 — 12
Blue	5 — 0	Orange	20 — 16
Cyan	8 — 5	Red.	23 — 20
Green	12 — 8		

Densities of the Elements (g/mL)

1	2	3	4	5	6	7	8	9	10	11	12	13	14	15	16	17	18
H * 0.09																	He * 0.18
Li 0.5	Be 1.8											B 2.3	C 2.0	N * 1.25	O * 1.43	F * 1.70	Ne * 0.90
Na 1.0	Mg 1.7											Al 2.7	Si 2.3	P 1.8	S 2.1	Cl * 3.21	Ar * 1.78
K 0.9	Ca 1.6	Sc 3.0	Ti 4.5	V 6.1	Cr 7.2	Mn 7.3	Fe 7.8	Co 8.9	Ni 8.9	Cu 9.0	Zn 7.1	Ga 6.1	Ge 5.3	As 5.7	Se 4.8	Br * 7.59	Kr * 3.73
Rb 1.5	Sr 2.5	Y 4.5	Zr 6.5	Nb 8.5	Mo 6.8	Tc 11.5	Ru 12.4	Rh 12.4	Pd 12.0	Ag 10.5	Cd 8.7	In 7.3	Sn 5.7	Sb 6.7	Te 6.2	I 4.9	Xe * 5.89
Cs 1.9	Ba 3.5	La 6.2	Hf 13.3	Ta 16.6	W 19.3	Re 21.0	Os 22.6	Ir 22.4	Pt 21.5	Au 18.9	Hg 13.5	Tl 11.9	Pb 11.4	Bi 9.7	Po 9.3	At --	Rn * 9.73
Fr --	Ra 5.0	Ac 10.1	Unq --	Unp --	Unh --	Uns --	Uno --	Une --									

* density of gaseous phase in g/L

Lanthanides:	Ce 6.7	Pr 6.7	Nd 6.8	Pm 7.2	Sm 7.5	Eu 5.2	Gd 7.9	Tb 8.2	Dy 8.6	Ho 8.8	Er 9.1	Tm 9.3	Yb 6.9	Lu 9.8

Actinides:	Th 11.7	Pa 15.4	U 19.0	Np 20.1	Pu 19.8	Am 13.7	Cm 13.5	Bk 14	Cf --	Es --	Fm --	Md --	No --	Lr --

1. Which elements are the most dense?

_____

2. How variable are the densities of the lanthanides compared with the densities of the actinides?

_____

3. Which atomic groups tend to go from higher to lower densities reading from top to bottom? (Identify each group by its group number.)

_____

4. Which atomic groups tend to go from lower to higher densities reading from top to bottom?

_____

Conceptual Integrated Science — Third Edition

Chapter 11: Investigating Matter
The Submicroscopic

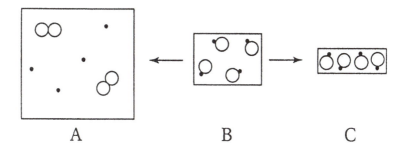

	A	B	C

1. How many molecules are shown in   A _____   B _____   C _____

2. How many atoms are shown in   A _____   B _____   C _____

3. Which represents a physical change?   B ⟶ A   B ⟶ C   (*circle one*)

4. Which represents a chemical change?   B ⟶ A   B ⟶ C   (*circle one*)

5. Which box(es) represent(s) a mixture?   A _____   B _____   C _____

6. Which box contains the most mass?   A _____   B _____   C _____

7. Which box is the coldest?   A _____   B _____   C _____

8. Which box contains the most air between molecules?   A _____   B _____   C _____

---

9. How many molecules are shown in   A _____   B _____   C _____

10. How many atoms are shown in   A _____   B _____   C _____

11. Which represents a physical change?   B ⟶ A   B ⟶ C   (*circle one*)

12. Which represents a chemical change?   B ⟶ A   B ⟶ C   (*circle one*)

13. Which box(es) represent(s) a mixture?   A _____   B _____   C _____

14. Which box contains the most mass?   A _____   B _____   C _____

15. Which should take longer?   B ⟶ A   B ⟶ C   (*circle one*)

16. Which box most likely contains ions?   A _____   B _____   C _____

Name _____ Date _____

Conceptual Integrated Science — Third Edition

Chapter 11: Investigating Matter
Physical and Chemical Changes

1. What distinguishes a chemical change from a physical change?

2. On the basis of observations alone, why is distinguishing a chemical change from a physical change not always so straightforward?

Try your hand at categorizing the following processes as either chemical or physical change. Some of these examples are debatable! Be sure to discuss your reasoning with your classmates or your instructor.

(circle one)

3. A cloud grows dark._____	chemical	physical
4. Leaves produce oxygen._____	chemical	physical
5. Food coloring is added to water. _____	chemical	physical
6. Tropical coral reef dies._____	chemical	physical
7. Dead coral reef is pounded by waves into beach sand. _____	chemical	physical
8. Oil and vinegar separate. _____	chemical	physical
9. Soda drink goes flat. _____	chemical	physical
10. Sick person develops a fever. _____	chemical	physical
11. Compost pit turns into mulch. _____	chemical	physical
12. A computer is turned on. _____	chemical	physical
13. An electrical short melts a computer's integrated circuits. _____	chemical	physical
14. A car battery runs down. _____	chemical	physical
15. A pencil is sharpened. _____	chemical	physical
16. Mascara is applied to eyelashes. _____	chemical	physical
17. Sunbather gets tan lying in the sun. _____	chemical	physical
18. Invisible ink turns visible upon heating. _____	chemical	physical
19. A light bulb burns out. _____	chemical	physical
20. Car engine consumes a tank of gasoline. _____	chemical	physical
21. B vitamins turn urine yellow. _____	chemical	physical

Chapter 12: Chemical Bonds and Mixtures

Losing Valence Electrons

The shell model described in Section 12.1 can be used to explain a wide variety of properties of atoms. Using the shell model, for example, we can explain how atoms within the same group tend to lose (or gain) the same number of electrons. Let's consider the case of three group 1 elements: lithium, sodium, and potassium. Look to a periodic table and find the nuclear charge of each of these atoms.

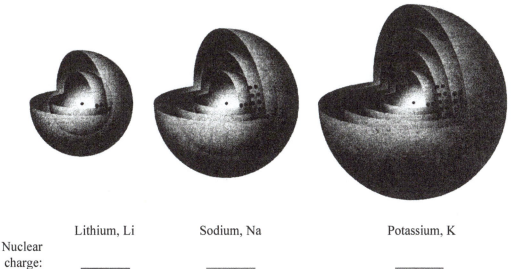

	Lithium, Li	Sodium, Na	Potassium, K
Nuclear charge:	_____	_____	_____
Number of inner shell electrons:	_____	_____	_____

How strongly the valence electron is held to the nucleus depends on the strength of the nuclear charge—the stronger the charge, the stronger the valence electron is held. There's more to it, however, because inner-shell electrons weaken the attraction outer-shell electrons have for the nucleus. The valence shell in lithium, for example, doesn't experience the full effect of three protons. Instead, it experiences a diminished nuclear charge of about +1. We get this by subtracting the number of inner-shell electrons from the actual nuclear charge. What do the valence electrons for sodium and potassium experience?

Diminished nuclear charge: _____   _____   _____

Question: Potassium has a nuclear charge many times greater than that of lithium. Why is it actually *easier* for a potassium atom to lose its valence electron than it is for a lithium atom to lose its valence electron?

Hint: Remember from Chapter 7 what happens to the electric force as distance is increased!

Conceptual Integrated Science ─── **Third Edition**

Chapter 12: Chemical Bonds and Mixtures

Drawing Shells

Atomic shells can be represented by a series of concentric circles as shown in your textbook. With a little effort, however, it's possible to show these shells in three dimensions. Grab a pencil and blank sheet of paper and follow the steps shown below. Practice makes perfect.

1. Lightly draw a diagonal guideline. Then, draw a series of seven semicircles. Note how the ends of the semicircles are not perpendicular to the guideline. Instead, they are parallel to the length of the page, as shown in Figure 1.

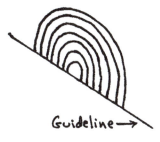

Guideline→

Figure 1

Figure 2

2. Connect the ends of each semicircle with those of another semicircle such that a series of concentric hearts are drawn. The ends of these new semicircles should be drawn perpendicular to the ends of the previously drawn semicircles, as shown in Figure 2.

3. Now the hard part. Draw a portion of a circle that connects the apex of the largest vertical and horizontal semicircles, as in Figure 3.

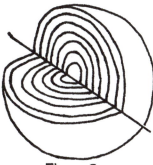

Figure 3

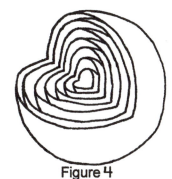

Figure 4

4. Now the fun part. Erase the pencil guideline then add the internal lines, as shown in Figure 4, that create a series of concentric shells.

You need not draw all the shells for each atom. Oxygen, for example, is nicely represented drawing only the first two inner shells, which are the only ones that contain electrons. Remember that these shells are not to be taken literally. Rather, they are a highly simplified view of how electrons tend to organize themselves within an atom. You should know that each shell represents a set of atomic orbitals of similar energy levels as shown in your textbook.

Conceptual Integrated Science Third Edition

Chapter 12: Chemical Bonds and Mixtures

Atomic Size

1. Complete the shells for the following atoms using arrows to represent electrons.

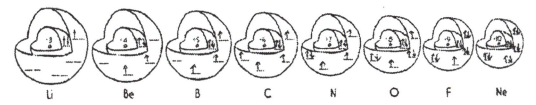

Li   Be   B   C   N   O   F   Ne

2. Neon, Ne, has many more electrons than lithium, Li, yet it is a much smaller atom. Why?

3. Draw the shell model for a sodium atom, Na (atomic number 11), adjacent to the neon atom in the box shown below. Use a pencil because you may need to erase.

a. Which should be larger, neon's first shell or sodium's first shell? Why? Did you represent this accurately within your drawing?

b. Which has a greater nuclear charge, Ne or Na?

c. Which is a larger atom, Ne or Na?

Ne   Na

4. Moving from left to right across the periodic table, what happens to the nuclear charge within atoms? What happens to atomic size?

5. Moving from top to bottom down the periodic table, what happens to the number of occupied shells? What happens to atomic size?

6. Where in the periodic table are the smallest atoms found? Where are the largest atoms found?

Conceptual Integrated Science — Third Edition

Chapter 12: Chemical Bonds and Mixtures
Effective Nuclear Charge

The magnitude of the nuclear charge sensed by an orbiting electron depends upon several factors, including the number of positively charged protons in the nucleus, the number of inner shell electrons shielding it from the nucleus, and its distance from the nucleus.

1. Place the proper number of electrons in each shell for carbon and silicon (use arrows to represent electrons).

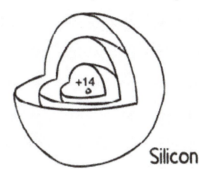

Carbon   Silicon

2. According to the shell model, which should experience the greater effective nuclear charge, an electron in

   a. carbon's 1st shell or silicon's 1st shell?   (circle one)

   b. carbon's 2nd shell or silicon's 2nd shell?   (circle one)

   c. carbon's 2nd shell or silicon's 3rd shell?   (circle one)

3. List the shells of carbon and silicon in order of decreasing effective nuclear charge.

4. Which should have the greater ionization energy, the carbon atom or the silicon atom? Defend your answer.

5. How many additional electrons are able to fit in the outermost shell of carbon? _____ Silicon? _____

6. Which should be stronger, a C-H bond or an Si-H bond? Defend your answer.

7. Which should be larger in size, the ion C^{4+} or the ion Si^{4+}? Why?

Conceptual Integrated Science ── Third Edition

Chapter 12: Chemical Bonds and Mixtures
Solutions

1. Use these terms to complete the following sentences. Some terms may be used more than once.

solution	solvent	solute
dissolve	concentrated	dilute
saturated	concentration	mole
molarity	solubility	soluble
insoluble	precipitate	

Sugar is _____ in water for the two can be mixed homogeneously to form a _____. The _____ of sugar in water is so great that _____ homogeneous mixtures are easily prepared. Sugar, however, is not infinitely _____ in water, for when too much of this _____ is added to water, which behaves as the _____, the solution becomes _____. At this point any additional sugar is _____ for it will not _____. If the temperature of a saturated sugar solution is lowered, the _____ of the sugar in water is also lowered. If some of the sugar comes out of solution, it is said to form a _____. If, however, the sugar remains in solution despite the decrease in solubility, then the solution is said to be supersaturated. Adding only a small amount of sugar to water results in a _____ solution. The _____ of this solution or any solution can be measured in terms of _____, which tells us the number of solute molecules per liter of solution. If there are 6.022×10^{23} molecules in 1 liter of solution, then the _____ of the solution is 1 _____ per liter.

2. Temperature has a variety of effects on the solubilities of various solutes. With some solutes, such as sugar, solubility increases with increasing temperature. With other solutes, such as sodium chloride (table salt), changing temperature has no significant effect. With some solutes, such as lithium sulfate (Li_2SO_4) the solubility actually decreases with increasing temperature.

a. Describe how you would prepare a supersaturated solution of lithium sulfate.

_____

b. How might you cause a saturated solution of lithium sulfate to form a precipitate?

_____

Name _____ Date _____

┌ Conceptual Integrated Science ── Third Edition

Chapter 12: Chemical Bonds and Mixtures
Pure Mathematics

Using a scientist's definition of *pure*, identify whether each of the following is 100% pure:

	100% pure?	
Freshly squeezed orange juice	Yes	No
Country air	Yes	No
Ocean water...	Yes	No
Fresh drinking water	Yes	No
Skim milk..	Yes	No
Stainless steel	Yes	No
A single water molecule.......................	Yes	No

A glass of water contains in the order of a trillion trillion (1×10^{24}) molecules. If the water in this were 99.9999% pure, you could calculate the percentage of impurities by subtracting from 100.0000%.

$$100.0000\% \text{ water} + \text{impurity molecules}$$
$$\underline{- \ 99.9999\% \text{ water molecules}}$$
$$0.0001\% \text{ impurity molecules}$$

Pull out your calculator and calculate the number of impurity molecules in the glass of water. Do this by finding 0.0001% of 1×10^{24}, which is the same as multiplying 1×10^{24} by 0.000001.

$$(1 \times 10^{24})(0.000001) = \underline{\qquad\qquad}$$

1. How many impurity molecules are there in a glass of water that's 99.9999% pure?

   a. 1000 (one thousand: 10^3)

   b. 1,000,000 (one million: 10^6)

   c. 1,000,000,000 (one billion: 10^9)

   d. 1,000,000,000,000,000,000 (one million trillion: 10^{18})

2. How does your answer make you feel about drinking water that is 99.9999 % free of some poison such as pesticide?

   _____

3. For every one impurity molecule, how many water molecules are there? (Divide the number of water molecules by the number of impurity molecules.)

   _____

4. Would you describe these impurity molecules within water that's 99.9999% pure as "rare" or "common"?

   _____

5. A friend argues that he or she doesn't drink tap water because it contains thousands of molecules of some impurity in each glass. How would you respond in defense of the water's purity, if it indeed does contain thousands of molecules of some impurity per glass?

   _____

   _____

─ Conceptual Integrated Science ── Third Edition

Chapter 12: Chemical Bonds and Mixtures
Chemical Bonds

1. On the basis of their positions in the periodic table, predict whether each pair of elements will form an ionic bond, covalent bond, or neither (atomic number in parenthesis).

a. Gold (79) and platinum (78) _____   f. Germanium (32) and arsenic (33) _____

b. Rubidium (37) and iodine (53) _____   g. Iron (26) and chromium (24) _____

c. Sulfur (16) and chlorine (17) _____   h. Chlorine (17) and iodine (53) _____

d. Sulfur (16) and magnesium (12) _____   i. Carbon (6) and bromine (35) _____

e. Calcium (20) and chlorine (17) _____   j. Barium (56) and astatine (85) _____

2. The most common ions of lithium, magnesium, aluminum, chlorine, oxygen, and nitrogen and their respective charges are as follows:

Positively Charged Ions	Negatively Charged Ions
Lithium ion: Li^{1+}	Chloride ion: Cl^{1-}
Barium ion: Ba^{2+}	Oxide ion: O^{2-}
Aluminum ion: Al^{3+}	Nitride ion: N^{3-}

Use this information to predict the chemical formulas for the following ionic compounds:

a. Lithium chloride: _____   d. Lithium oxide: _____   g. Lithium nitride: _____

b. Barium chloride: _____   e. Barium oxide: _____   h. Barium nitride: _____

c. Aluminum chloride: _____   f. Aluminum oxide: _____   i. Aluminum nitride: _____

j. How are elements that form positive ions grouped in the periodic table relative to elements that form negative ions? _____

3. Specify whether the following chemical structures are polar or nonpolar:

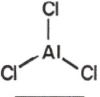

Conceptual Integrated Science
Third Edition

Chapter 12: Chemical Bonds and Mixtures
Shells and the Covalent Bond

When atoms bond covalently, their atomic shells overlap so that shared electrons can occupy both shells at the same time.

Nonbonded hydrogen atoms

 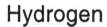 

Hydrogen   Hydrogen

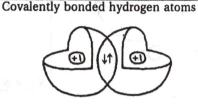

Covalently bonded hydrogen atoms

Molecular Hydrogen

Formula: H_2

Fill each shell model shown below with enough electrons to make each atom electrically neutral. Use arrows to represent electrons. Within the box draw a sketch showing how the two atoms bond covalently. Draw hydrogen shells more than once when necessary so that no electrons remain unpaired. Write the name and chemical formula for each compound.

A.

Hydrogen   Carbon

Name of Compound: _____   Formula:

B.

Hydrogen   Nitrogen

Name of Compound: _____   Formula:

Shells and the Covalent Bond—continued

C.

Hydrogen   Oxygen

Name of Compound: Formula:

D.

Hydrogen   Fluorine

Name of Compound: Formula:

E.

Hydrogen   Neon

Name of Compound: Formula:

1. Note the relative positions of carbon, nitrogen, oxygen, fluorine, and neon in the periodic table. How does this relate to the number of times each of these elements is able to bond with hydrogen?

_____

_____

2. How many times is the element boron (atomic number 5) able to bond with hydrogen? Use the shell model to help you with your answer.

_____

_____

Conceptual Integrated Science
Third Edition

Chapter 12: Chemical Bonds and Mixtures
Bond Polarity

Pretend you are one of two electrons being shared by a hydrogen atom and a fluorine atom. Say, for the moment, you are centrally located between the two nuclei. You find that both nuclei are attracted to you. Hence, because of your presence, the two nuclei are held together.

You are here.

H : F

1. Why are the nuclei of these atoms attracted to you?_____

2. What type of chemical bonding is this?_____

You are held within hydrogen's 1st shell and at the same time within fluorine's 2nd shell. Draw a sketch using the shell models below to show how this is possible. Represent yourself and all other electrons using arrows. Note your particular location with a circle.

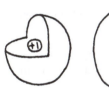

Hydrogen   Fluorine

Your Sketch

According to the laws of physics, if the nuclei are both attracted to you, then you are attracted to both of the nuclei.

3. You are pulled toward the hydrogen nucleus, which has a positive charge. How strong is this charge from your point of view—what is its *electronegativity?*_____

4. You are also attracted to the fluorine nucleus. What is its electronegativity?_____

You are being shared by the hydrogen and fluorine nuclei. But as a moving electron you have some choice as to your location.

5. Consider the electronegativities you experience from both nuclei. Which nucleus would you tend to be closest to? _____

Name _____ Date _____

Conceptual Integrated Science — Third Edition

Bond Polarity—continued

Stop pretending you are an electron and observe the hydrogen-fluorine bond from outside the hydrogen fluoride molecule. Bonding electrons tend to congregate to one side because of the differences in effective nuclear charges. This makes one side slightly negative in character and the opposite side slightly positive. Indicate this on the following structure for hydrogen fluoride using the symbols δ– and δ+

H ⦂ F

By convention, bonding electrons are not shown. Instead, a line is simply drawn connecting the two bonded atoms. Indicate the slightly negative and positive ends.

H — F

6. Would you describe hydrogen fluoride as a polar or nonpolar molecule?_____

7. If two hydrogen fluoride molecules were thrown together, would they stick or repel? (Hint: What happens when you throw two small magnets together?)_____

8. Place bonds between the hydrogen and fluorine atoms to show many hydrogen fluoride molecules grouped together. Each element should be bonded only once. Circle each molecule and indicate the slightly negative and slightly positive ends.

H	F	H	F	H	F	H	F	H	F
F	H	F	H	F	H	F	H	F	H
H	F	H	F	H	F	H	F	H	F
F	H	F	H	F	H	F	H	F	H
H	F	H	F	H	F	H	F	H	F
F	H	F	H	F	H	F	H	F	H

— Conceptual Integrated Science — Third Edition

Chapter 12: Chemical Bonds and Mixtures
Atoms to Molecules

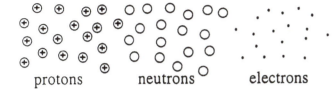

protons   neutrons   electrons

SUBATOMIC PARTICLES

Subatomic particles are the fundamental building blocks of all _____.

hydrogen atom

hydrogen atom

oxygen atom

oxygen atom   hydrogen atom   hydrogen atom

ATOMS

An atom is a group of _____ _____ held tightly together. An oxygen atom is a group of 8 _____, 8 _____, and 8 _____. A hydrogen atom is a group of only 1 _____ and 1 _____.

water molecule   water molecule

MOLECULES

A _____ is a group of atoms held tightly together. A water _____ consists of 2 _____ atoms and 1 _____ atom.

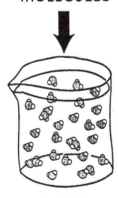

WATER

Water is a material made up of billions upon billions of water _____. The physical properties of water are based upon how these water _____ interact with one another. The electronic attraction between _____ is one of the major topics of Chapter 12.

Conceptual Integrated Science Third Edition

Chapter 12: Chemical Bonds and Mixtures
Electron-Dot Structures

Electron-dot structures tell us how atoms tend to bond with other atoms. Carbon's electron-dot structure, for example, shows 4 unpaired valence electrons. Each hydrogen atom has 1 unpaired electron. Unpaired electrons from different atoms can pair up, resulting in a bond. For carbon and hydrogen, we have the following, which creates the molecule methane, CH_4:

Use this idea to draw the electron-dot structures of the products indicated by name.

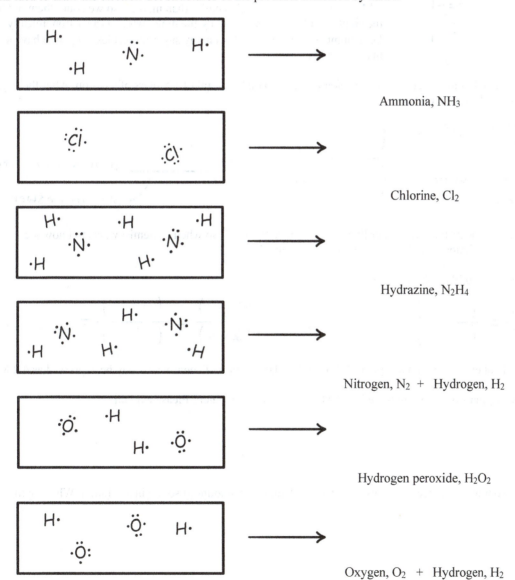

Ammonia, NH_3

Chlorine, Cl_2

Hydrazine, N_2H_4

Nitrogen, N_2 + Hydrogen, H_2

Hydrogen peroxide, H_2O_2

Oxygen, O_2 + Hydrogen, H_2

Name _____   Date _____

Conceptual Integrated Science — Third Edition

Chapter 12: Chemical Bonds and Mixtures
Molarity

Concentration is a measure of the amount of solute within a solution. For the solute, chemists often go by a count of the number of solute particles. For the solution, they usually consider the volume of the solution in units of liters. Note, the volume of solution is the combined volume of the solute and the solvent.

   For example, a chemist drops 5 marbles into a beaker of water to make a total of 0.5 liters. The concentration would be:

$$\frac{5 \text{ marbles}}{0.5 \text{ } \mathcal{L}} = \underline{\hspace{2cm}} \text{ marbles per liter}$$

plug in answer

Of course, marbles don't dissolve in water, but they do serve to illustrate the point that it's not just the solvent that contributes to the volume of a solution.

   Molecules are much, much smaller than marbles so we count them not by the single, not by the dozen, but by the mole, which is an astronomically large number: 6.02×10^{23}. This many sugar molecules happen to have a mass of 342 grams.

A chemist drops 684 grams of sugar into a beaker of water to make a total of 1.5 liters of solution. After the sugar dissovles, what is the concentration?

amount of sugar (in moles) →

volume of solution → (in $\mathcal{L}$)

$$\frac{(\quad\quad)}{(\quad\quad)} = \underline{\hspace{2cm}} \text{ moles per liter}$$

plug in answer

In the language of chemistry, "moles per liter" is also called "molar." So when a chemist wants to know the concentration of a solution, she'll ask, "What's the molarity?"

What is the molarity of the following solutions?

$$\frac{(171 \text{ g Sugar})}{(3 \text{ } \mathcal{L} \text{ soln})} = \frac{(\quad\quad)}{(\quad\quad)} = \underline{\hspace{1.5cm}} \text{ molar}$$ $$\frac{(342 \text{ g Sugar})}{(1 \text{ } \mathcal{L} \text{ soln})} = \frac{(\quad\quad)}{(\quad\quad)} = \underline{\hspace{1.5cm}} \text{ molar}$$

Molar is usually abbreviated with a capital M. So a 4.2 molar solution of sugar water may be expressed as 4.2 M.

1. So how many grams of sugar are there in a 3 M solution of sugar water? Please explain.

2. There are two things you need to know in order to calculate the amount of solute in a solution. What are these two things?

Conceptual Integrated Science
Third Edition

Chapter 12: Chemical Bonds and Mixtures
Polyatomic Ions

Sometimes a molecule can lose or gain a proton (hydrogen ion) to form what we call a polyatomic ion:

| Phosphoric acid (molecule) | Phosphate ion (polyatomic ion) | Ammonia (molecule) | Ammonium ion (polyatomic ion) |

Table of common polyatomic ions

NAME	FORMULA	NAME	FORMULA
Ammonium ion	NH_4^+	Hydroxide ion	OH^-
Bicarbonate ion	HCO_3^-	Nitrate ion	NO_3^-
Carbonate ion	CO_3^{2-}	Phosphate ion	PO_4^{3-}
Cyanide ion	CN^-	Sulfate ion	SO_4^{2-}

When it comes to naming compounds, a polyatomic ion is treated as a single unit. Positively charged ions are listed first followed by the negatively charged ions, but we don't include the word "ion." For example, below is the formula for ammonium phosphate. Note how we need three (1+) ammoniums to balance a single (3–) phosphate.

positively charged ion   negatively charged ion

$$(NH_4)_3 PO_4$$

Use the table of common polyatomic ions to deduce the formula for the following compounds:

Ammonium sulfate _____   Potassium cyanide _____

Sodium sulfate _____   Calcium phosphate _____

Sodium hydroxide _____   Aluminum hydroxide _____

Hydrogen hydroxide _____   Aluminum sulfate _____

Name the following structures and write their formula on the basis of the polyatomic ions they contain:

Name: _____   Name: _____

Formula: _____   Formula: _____

Name: _____   Name: _____

Formula: _____   Formula: _____

Conceptual Integrated Science
Third Edition

Chapter 13: Chemical Reactions
Balancing Chemical Equations

In a balanced chemical equation the number of times each element appears as a reactant is equal to the number of times it appears as a product. For example,

$$2 \ H_2 \ + \ O_2 \ ---> \ 2 \ H_2O$$

Recall that *coefficients* (the integer appearing before the chemical formula) indicate the number of times each chemical formula is to be counted and *subscripts* indicate when a particular element occurs more than once within the formula.

Check whether the following chemical equations are balanced:

$3 \ NO \ ---> \ N_2O \ + \ NO_2$ ☐ balanced ☐ unbalanced

$SiO_2 \ + \ 4 \ HF \ ---> \ SiF_4 \ + \ 2 \ H_2O$ ☐ balanced ☐ unbalanced

$4 \ NH_3 \ + \ 5 \ O_2 \ ---> \ 4 \ NO \ + \ 6 \ H_2O$ ☐ balanced ☐ unbalanced

Unbalanced equations are balanced by changing the coefficients. Subscripts, however, should never be changed because this changes the chemical's identity—H_2O is water, but H_2O_2 is hydrogen peroxide! The following steps may help guide you:

1. Focus on balancing only one element at a time. Start with the left-most element and modify the coefficients such that this element appears on both sides of the arrow the same number of times.

2. Move to the next element and modify the coefficients so as to balance this element. Do not worry if you incidentally unbalance the previous element. You will come back to it in subsequent steps.

3. Continue from left to right, balancing each element individually.

4. Repeat steps 1–3 until all elements are balanced.

Use the above methodology to balance the following chemical equations:

____N_2O ---> ____N_2 + ____O_2

____$NaClO_3$ ---> ____$NaCl$ + ____O_2

____$MnCl_2$ + ____Al ---> ____Mn + ____$AlCl_3$

____K + ____H_2O ---> ____H_2 + ____KOH

____Al_2O_3 + ____C ---> ____Al + ____CO_2

____NH_3 + ____F_2 ---> ____NH_4F + ____NF_3

This is just one of the many methods that chemists have developed to balance chemical equations.

Knowing how to balance a chemical equation is a useful technique, but understanding why a chemical equation needs to be balanced in the first place is far more important.

Conceptual Integrated Science — Third Edition

Chapter 13: Chemical Reactions

Exothermic and Endothermic Reactions

During a chemical reaction atoms are neither created nor destroyed. Instead, atoms rearrange—they change partners. This rearrangement of atoms necessarily involves the input and output of energy. First, energy must be supplied to break chemical bonds that hold atoms together. Separated atoms then form new chemical bonds, which involves the release of energy. In an **exothermic** reaction more energy is released than is consumed. Conversely, in an **endothermic** reaction more energy is consumed than is released.

Table 1 Bond Energies

Bond	Bond Energy*	Bond	Bond Energy*
H—H	436	Cl—Cl	243
H—C	414	N—N	159
H—N	389	O=O	498
H—O	464	O=C	803
H—Cl	431	N≡N	946

*In kJ/mol

Table 1 shows bond energies—the amount of energy required to break a chemical bond, and also the amount of energy released when a bond is formed. Use these bond energies to determine whether the following chemical reactions are exothermic or endothermic:

Total Amount of Energy
Required to Break Bonds
_____ kJ/mol

Total Amount of Energy
Released Upon Bond Formation
_____ kJ/mol

Net Energy Change of Reaction: _____ kJ/mole   (absorbed) (released)
circle one

1. Is this reaction exothermic or endothermic?

2. Write the balanced equation for this reaction using chemical formulas and coefficients. If it is exothermic, write "Energy" as a product. If it is endothermic, write "Energy" as a reactant.

Conceptual Integrated Science — Third Edition

Exothermic and Endothermic Reactions—continued

H—C—H + O=O → O=C=O + H—O—H

Methane   Oxygen   Carbon   Water
                          Dioxide

Total Amount of Energy   Total Amount of Energy
Required to Break Bonds   Released Upon Bond Formation
_____kJ/mol   _____kJ/mol

Net Energy Change of Reaction: _____kJ/mole   (absorbed /released)
                                                      circle one

3. Is this reaction exothermic or endothermic?

4. Write the balanced equation for this reaction using chemical formulas and coefficients. If it is exothermic write "Energy" as a product. If it is endothermic write "Energy" as a reactant.

N≡N + H—H → H—N—N—H

Nitrogen   Hydrogen   Hydrazine

Total Amount of Energy   Total Amount of Energy
Required to Break Bonds   Released Upon Bond Formation
_____ kJ/mol   _____ kJ/mol

Net Energy Change of Reaction: _____kJ/mole   (absorbed/released)
                                                      circle one

5. Is this reaction exothermic or endothermic?

6. Write the balanced equation for this reaction using chemical formulas and coefficients. If it is exothermic, write "Energy" as a product. If it is endothermic, write "Energy" as a reactant.

—Conceptual Integrated Science— Third Edition

Chapter 13: Chemical Reactions

Donating and Accepting Hydrogen Ions

A chemical reaction that involves the transfer of a hydrogen ion from one molecule to another is classified as an acid–base reaction. The molecule that donates the hydrogen ion behaves as an acid. The molecule that accepts the hydrogen ion behaves as a base.

On paper, the acid–base process can be depicted through a series of frames:

Frame 1

ammonium   hydroxide
ion   ion

Ammonium and hydroxide ions in close proximity

Frame 2

Bond is broken between the nitrogen and a hydrogen of the ammonium ion. The two electrons of the broken bond stay with the nitrogen leaving the hydrogen with a positive charge.

Frame 3

The hydrogen ion migrates to the hydroxide ion.

Frame 4

The hydrogen ion bonds with the hydroxide ion to form a water molecule.

In equation form we abbreviate this process by only showing the before and after:

frame 1   frame 4

137

Conceptual Integrated Science
Third Edition

Donating and Accepting Hydrogen Ions—continued

We see from the previous reaction that because the ammonium ion donated a hydrogen ion, it behaved as an acid. Conversely, the hydroxide ion by accepting a hydrogen ion behaved as a base. How do the ammonia and water molecules behave during the reverse process?

| acid | base | ammonia | water |

Identify the following molecules as behaving as an acid or a base:

— Conceptual Integrated Science — Third Edition

Chapter 13: Chemical Reactions

Loss and Gain of Electrons

A chemical reaction that involves the transfer of an electron is classified as an oxidation–reduction reaction. Oxidation is the process of losing electrons, while reduction is the process of gaining them. Any chemical that causes another chemical to lose electrons (become oxidized) is called an *oxidizing agent*. Conversely, any chemical that causes another chemical to gain electrons is called a *reducing agent*.

1. What is the relationship between an atom's ability to behave as an oxidizing agent and its electron affinity?

2. Relative to the periodic table, which elements tend to behave as strong oxidizing agents?

3. Why don't the noble gases behave as oxidizing agents?

4. How is it that an oxidizing agent is itself reduced?

5. Specify whether each reactant is about to be oxidized or reduced.

$$2\,K \;+\; H_2O \;\longrightarrow\; 2\,K^+ \;+\; {}^-OH$$
_____   _____

$$2\,Mg \;+\; O_2 \;\longrightarrow\; 2\,Mg^{2+}O^{2-}$$
_____   _____

$$2\,Na \;+\; Cl_2 \;\longrightarrow\; 2\,Na^+Cl^-$$
_____   _____

$$CH_4 \;+\; 2\,O_2 \;\longrightarrow\; O{=}C{=}O \;+\; {}_H{\nearrow}^{O-H}$$
_____   _____

6. Which oxygen atom enjoys a greater negative charge?

— this one   — that one

$O{=}O$ or $H{-}O{\diagdown}_H$ (*circle one*)

7. Relate your answer to Question 6 to how it is that O_2 is reduced upon reacting with CH_4 to form carbon dioxide and water.

─ Conceptual Integrated Science ─ Third Edition

Chapter 13: Chemical Reactions
Ocean Acidification

Because carbon dioxide is a non polar gas, you might expect that it does not mix readily with water, which is a polar liquid. However, as soon as carbon dioxide enters water, it reacts to form a new substance called carbonic acid, which has much better water solubility. Because of this chemistry, water is able to absorb relatively large quanitities of carbon dioxide.

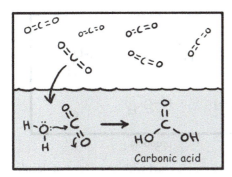

1. What happens to the pH of water as more CO_2 molecules are drawn into it?

Carbonic acid, in turn, can transform back into carbon dioxide as shown below. In general, the higher the temperature, the more readily this happens.

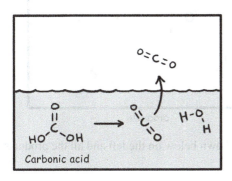

2. Why does seltzer water have such a low pH? (around 3.5!)

3. Why does warm seltzer water go flat faster than cold seltzer water?

There's even more chemistry when it comes to our oceans, which are alkaline because of dissolved compounds such as calcium carbonate, $CaCO_3$.

4. Does the alkalinity of the oceans help or hinder the absorption of carbon dioxide? Please explain.

5. What happens to the pH of the oceans as they absorb increasing amounts of carbon dioxide?

6. How is a coral reef connected to a coal-fired power plant?

7. If the oceans can absorb carbon dioxide, why have the atmospheric levels of carbon dioxide been rising so rapidly over the past century?

Conceptual Integrated Science — Third Edition

Chapter 13: Chemical Reactions
Fuel Cells

The reaction of 4 hydroxide ions, HO^-, with 2 hydrogen molecules, H_2, creates 4 water molecules, H_2O, plus 4 electrons, e^-:

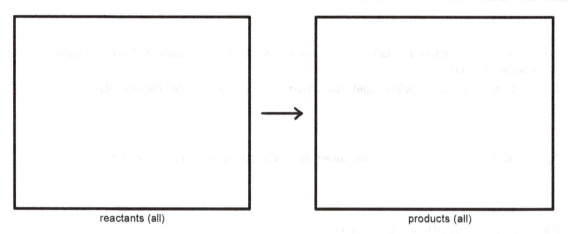

reactants   products

In a fuel cell this can be coupled with a second reaction in which 4 electrons, e^-, react with an oxygen molecule, O_2, and 2 water molecules, H_2O, to form 4 hydroxide ions, HO^-:

reactants   products

Combine all the reactants from the above two equations into the box shown below on the left and all the products from above into the box on the right to create a single big equation:

reactants (all)   products (all)

Each of these reactants and products appears at some time during the operation of this type of hydrogen–oxygen fuel cell. But what is the net result? To find out, cross out each reactant on the left also appearing as a product on the right. For example, the two reactant water molecules should be crossed out along with two of the product water molecules.

1. Write the net equation you come up with after crossing out duplicate chemicals.

┌─ **Conceptual Integrated Science** ─── **Third Edition**

2. Is this reaction endothermic or exothermic? How does the light bulb become lit?

3. Use the following graphic to explain how the hydroxide ions you crossed out in the products eventually become hydroxide ions of the reactants.

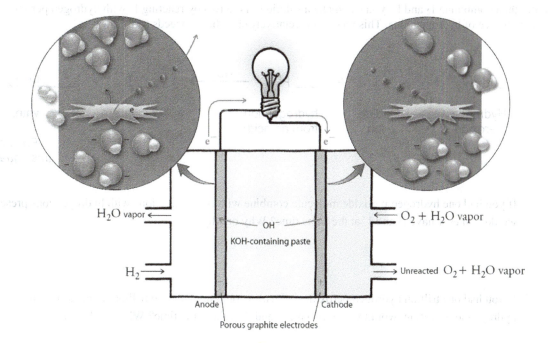

4. The anode provides electrons to the cathode. Will these electrons keep flowing if the oxygen supply is cut off? Please explain.

5. The cathode collects electrons from the anode. Will this continue if the hydrogen supply is cut off? Please explain.

Conceptual Integrated Science
Third Edition

Chapter 13: Chemical Reactions
Iodine Clock Reaction

Combine iodine, I_2, with iodide, I^-, and they will react to form the triiodide ion, I_3^-, which quickly binds with starch to form a dark blue color when in solution.

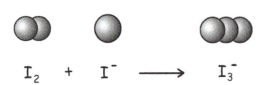

$$I_2 \ + \ I^- \ \longrightarrow \ I_3^-$$

Triiodide-starch complex
(dark blue)

Rather than combining I_2 and I^-, you can make a solution of the two by reacting I^- with hydrogen peroxide, H_2O_2, in the presence of hydrogen ions. This reaction is relatively slow but proceeds as follows:

$$\underline{\quad} H_2O_2 \ + \ \underline{\quad} I^- \ + \ \underline{\quad} H^+ \ \xrightarrow{\text{slow}} \ \underline{\quad} I_2 \ + \ \underline{\quad} H_2O$$

| Hydrogen peroxide | iodide ion | hydrogen ion (from an acid) | iodine | water |

Balance this equation

1. If you had one hydrogen peroxide molecule combine with two iodine ions with hydrogen ions present, would you ever have I_2 and I^- at the same time? Why or why not?

2. If you had one trillion hydrogen peroxide molecules combine with two trillion iodine ions with hydrogen ions present, would you ever have I_2 and I^- at the same time? Why or why not?

3. How does the slowness of this reaction play a role?

4. The reaction of H_2O_2 with I^- to form I_2 is quite slow. How might the rate of this reaction be increased?

─Conceptual Integrated Science ─── Third Edition

Vitamin C causes iodine, I_2, to break down rapidly into iodide ions as per the following reaction:

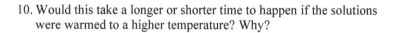

| Vitamin C | Iodine | | Dehydroxyascorbic acid | iodide ions | hydrogen ions |

5. A solution of one trillion hydrogen peroxide molecules and two trillion iodide ions (with plenty of hydrogen ions present) forms plenty of iodine, I_2, molecules. But what happens to these I_2 molecules if vitamin C is also in the solution?

6. Do I_2 and I^- ever coexist, practically speaking, with the vitamin C present? Why or why not?

7. As shown in the chemical equation at the top of the page, the vitamin C molecule is no longer a vitamin C molecule after it reacts with I_2. For the above solution, what would eventually happen if only 500 billion vitamin C molecules were present?

8. At what point will I_2 molecules and I^- ions finally be able to coexist in significant amounts?

9. If starch is also present, what happens to the color of the solution at that point?

10. Would this take a longer or shorter time to happen if the solutions were warmed to a higher temperature? Why?

11. Water disinfected with iodine, I_2, can taste a bit like iodine. Yuk! What might be added to this I_2 disinfected water to remove some of this taste?

Name _____ Date _____

Conceptual Integrated Science
Third Edition

Chapter 14: Organic Compounds
Structures of Organic Compounds

1. What are the chemical formulas for the following structures?

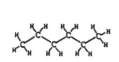   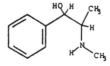

Formula: _____   _____   _____   _____

2. How many covalent bonds is carbon able to form? _____

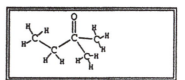

3. What is wrong with the structure shown in the box at right?

_____

4. a. Draw a hydrocarbon that contains 4 carbon atoms.

   b. Redraw your structure and transform it into an amine.

   c. Transform your amine into an amide. You may need to relocate the nitrogen.

   d. Redraw your amide, transforming it into a carboxylic acid.

   e. Redraw your carboxylic acid, transforming it into an alcohol.

   f. Rearrange the carbons of your alcohol to make an ether.

─ **Conceptual Integrated Science** ─ Third Edition

Chapter 14: Organic Compounds

Polymers

1. Circle the monomers that may be useful for forming an addition polymer and draw a box around the ones that may be useful for forming a condensation polymer.

2. Which type of polymer always weighs less than the sum of its parts? Why?

3. Would a material with the following arrangement of polymer molecules have a relatively high or low melting point? Why?

Conceptual Integrated Science Third Edition

Chapter 15: The Basic Unit of Life—The Cell
Prokaryotic Cells and Eukaryotic Cells

1. State whether the following features are found in prokaryotic cells, eukaryotic cells, or both.

   a. nucleic acids   _____

   b. cell membrane   _____

   c. nucleus   _____

   d. organelles   _____

   e. mitochondria   _____

   f. chloroplasts   _____

   g. circular chromosome   _____

   h. cytoplasm   _____

2. State whether these are prokaryotes or eukaryotes.

Human

Bacteria

Tree

a. _____   b. _____   c. _____

Bird

Mushroom

d. _____   e. _____

Conceptual Integrated Science — Third Edition

Chapter 15: The Basic Unit of Life—The Cell
Eukaryotic Organelles

1. Label the organelles in the animal cell with the following terms:

   Cytoskeleton   Ribosomes

   Golgi apparatus   Rough endoplasmic reticulum

   Lysosome   Smooth endoplasmic reticulum

   Mitochondrion

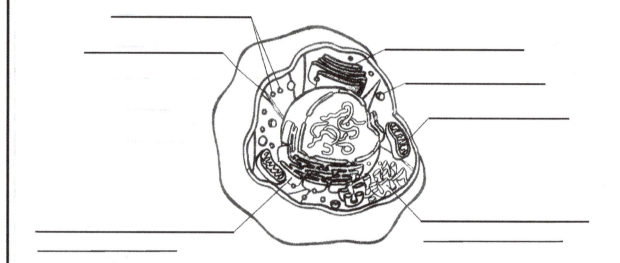

2. Match the organelles with their functions (a through h).

   a. Contains ribosomes for building certain   _____ Ribosome
      types of proteins   _____ Rough endoplasmic reticulum

   b. Breaks down organic molecules to   _____ Smooth endoplasmic reticulum
      obtain energy   _____ Golgi apparatus

   c. In plant cells, captures energy from   _____ Lysosome
      sunlight to build organic molecules   _____ Mitochondrion

   d. Receives products from the endoplasmic   _____ Chloroplast
      reticulum and packages them for transport   _____ Cytoskeleton

   e. Helps cell hold its shape

   f. Builds membranes

   g. Builds proteins

   h. Breaks down organic material

Conceptual Integrated Science — **Third Edition**

Chapter 15: The Basic Unit of Life—The Cell
The Cell Membrane

1. Label the three components of the cell membrane in the figure.

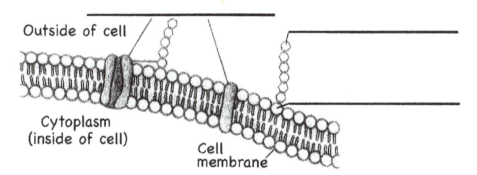

Outside of cell

Cytoplasm
(inside of cell)

Cell
membrane

2. a. Label the hydrophilic part and the hydrophobic part of the phospholipid.

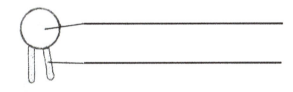

   b. Explain how the phospholipids are arranged in the cell membrane.

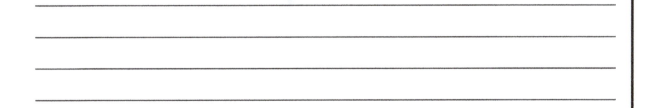

   _____

   _____

   _____

   _____

   _____

3. What do the membrane proteins do?

   _____

   _____

4. What do the short carbohydrates do?

   _____

Conceptual Integrated Science — Third Edition

Chapter 15: The Basic Unit of Life—The Cell

Diffusion and Osmosis

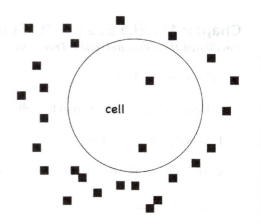

1. The molecules represented by squares move across the cell membrane through diffusion in the diagram on the right.

   Will there be a net movement of these molecules into the cell or out of the cell? Why?

Remember that these molecules move across the cell membrane through diffusion.

_____

_____

_____

_____

2. The diffusion of water has a special name.

   It is called _____.

   In the figure on the right, a membrane allows water to move freely between two compartments. The dark circles represent solute molecules, which are not able to move between the two compartments. Will water flow to the left or to the right? Why?

_____

_____

_____

_____

_____

In diffusion, molecules move from where they are more crowded to where they are less crowded.

Conceptual Integrated Science—Third Edition

Chapter 15: The Basic Unit of Life—The Cell
Facilitated Diffusion and Active Transport

1. Which of these describe _____, and which describe _____?

    a. Does not require energy from the cell _____

    b. Requires energy from the cell _____

    c. Moves molecules from a region of low concentration to a region of high concentration

    _____

    d. Moves molecules from a region of high concentration to a region of low concentration

    _____

2. Which of the following shows _____, and which shows _____?

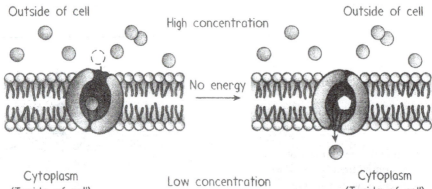

a. _____

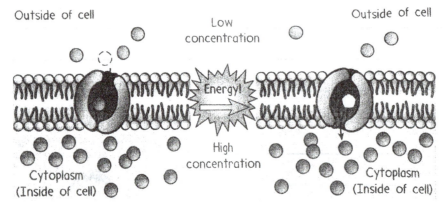

b. _____

Name _____ Date _____

Conceptual Integrated Science — Third Edition

Chapter 15: The Basic Unit of Life—The Cell

Cell Division

1. What is the function of cell division for single-celled organisms ?

    _____

2. What are some functions of cell division for multicellular organisms ?

    _____

3. What happens during each stage of the cell cycle ?

    Gap 1 _____

    Synthesis _____

    Gap 2 _____

    Mitosis and cytokinesis _____

4. Mitosis is divided into a number of phases. Describe what happens during each of the phases listed below.

    Prophase _____

    Metaphase _____

    Anaphase _____

    Telophase _____

Conceptual Integrated Science Third Edition

Chapter 15: The Basic Unit of Life—The Cell

Photosynthesis

1. During photosynthesis, one kind of energy, _____

   is converted into _____.

2. The chemical reaction for photosynthesis is

   _____ + _____ + _____ →

   _____ + _____

3. In what part of a plant cell does photosynthesis take place? _____

4. Why are plants green?

   _____

   _____

5. Explain why life as we know it would be impossible without photosynthesis.

   _____

   _____

   _____

   _____

   _____

   _____

Conceptual Integrated Science
Third Edition

Chapter 15: The Basic Unit of Life—The Cell
Cellular Respiration

1. The chemical reaction for cellular respiration is

   _____ + _____ + _____ →

   _____ + _____ + _____

2. Cells use cellular respiration to produce ATP. How do cells obtain energy from ATP?

   _____

   _____

   _____

Circle the correct answers.

3. Which of the processes requires oxygen?

   (Glycolysis)

   (Krebs cycle and electron transport)

   (Alcoholic fermentation)

   (Lactic acid fermentation)

4. During cellular respiration, most of the ATP is made during

   (glycolysis)   (Krebs cycle)   (electron transport).

Conceptual Integrated Science Third Edition

Chapter 16: Genetics
DNA Replication

1. The following is a piece of DNA:

```
AT
GC
CG
TA
TA
AT
CG
CG
GC
TA
AT
CG
GC
```

The strand is unwound, so that the DNA can be replicated. Fill in the nucleotides on the new strands.

—Conceptual Integrated Science Third Edition

Chapter 16: Genetics
Transcription and Translation

1. The figure below shows how information from DNA is used to build a protein. Write the names of the appropriate processes above the arrows.

 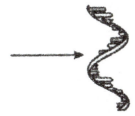 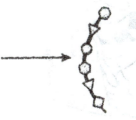

DNA   RNA   Protein

2. Transcription takes place in the cell's _____.

   During transcription, DNA is used to make a molecule of _____.

3. If the following strand of DNA is transcribed, what are the nucleotides found on the transcript?

   _____

   A T G G T C A T A C G T A C A A T G

4. Translation takes place in the cell's _____. Translation is performed by

   organelles called _____.

5. Divide the transcript from your answer to Question 3 into codons. Then, figure out the sequence of amino acids assembled in the ribosome. Use the genetic code table on the next page.

   Transcript   _____

   Codons   _____

   Amino acids   _____

Chapter 16: Genetics
Transcription and Translation—continued

A secret code!

Second base

	U	C	A	G	
U	UUU ⎤ Phenylalanine UUC ⎦ (Phe) UUA ⎤ Leucine UUG ⎦ (Leu)	UCU ⎤ UCC ⎥ Serine UCA ⎥ (Ser) UCG ⎦	UAU ⎤ Tyrosine UAC ⎦ (Tyr) UAA Stop UAG Stop	UGU ⎤ Cysteine UGC ⎦ (Cys) UGA Stop UGG Tryptophan (Trp)	U C A G
C	CUU ⎤ CUC ⎥ Leucine CUA ⎥ (Leu) CUG ⎦	CCU ⎤ CCC ⎥ Proline CCA ⎥ (Pro) CCG ⎦	CAU ⎤ Histidine CAC ⎦ (His) CAA ⎤ Glutamine CAG ⎦ (Gln)	CGU ⎤ CGC ⎥ Arginine CGA ⎥ (Arg) CGG ⎦	U C A G
A	AUU ⎤ AUC ⎥ Isoleucine AUA ⎥ (Ile) AUG Met or start	ACU ⎤ ACC ⎥ Threonine ACA ⎥ (Thr) ACG ⎦	AAU ⎤ Asparagine AAC ⎦ (Asn) AAA ⎤ Lysine AAG ⎦ (Lys)	AGU ⎤ Serine AGC ⎦ (Ser) AGA ⎤ Arginine AGG ⎦ (Arg)	U C A G
G	GUU ⎤ GUC ⎥ Valine GUA ⎥ (Val) GUG ⎦	GCU ⎤ GCC ⎥ Alanine GCA ⎥ (Ala) GCG ⎦	GAU ⎤ Aspartic GAC ⎦ acid(Asp) GAA ⎤ Glutamic GAG ⎦ acid (Glu)	GGU ⎤ GGC ⎥ Glycine GGA ⎥ (Gly) GGG ⎦	U C A G

First base · Third base

Name _____ Date _____

Conceptual Integrated Science
Third Edition

Chapter 16: Genetics
Genetic Mutations

1. Define the following terms:

   Genetic mutation _____

   _____

   Point mutation _____

   _____

   Nonsense mutation _____

   _____

   Frameshift mutation _____

   _____

2. Translate the following mRNA sequence into amino acids. You can use the genetic code table on the next page.

AAU	GUC	CCG	ACC	AAA	GCU
_____	_____	_____	_____	_____	_____

3. What point mutation in the sequence above could cause the substitution of the amino acid serine for asparagine?

   _____

   _____

4. How could a change in a single nucleotide in the sequence above result in a nonsense mutation?

   _____

   _____

   _____

5. The insertion or deletion of one or two nucleotides causes a frameshift mutation. Why doesn't the insertion or deletion of three nucleotides cause a frameshift mutation as well?

   _____

   _____

   _____

   _____

   _____

Second base

		U	C	A	G	
First base	**U**	UUU ⎫ Phenylalanine UUC ⎭ (Phe) UUA ⎫ Leucine UUG ⎭ (Leu)	UCU ⎫ UCC ⎬ Serine UCA ⎪ (Ser) UCG ⎭	UAU ⎫ Tyrosine UAC ⎭ (Tyr) UAA Stop UAG Stop	UGU ⎫ Cysteine UGC ⎭ (Cys) UGA Stop UGG Tryptophan (Trp)	U C A G
	C	CUU ⎫ CUC ⎬ Leucine CUA ⎪ (Leu) CUG ⎭	CCU ⎫ CCC ⎬ Proline CCA ⎪ (Pro) CCG ⎭	CAU ⎫ Histidine CAC ⎭ (His) CAA ⎫ Glutamine CAG ⎭ (Gln)	CGU ⎫ CGC ⎬ Arginine CGA ⎪ (Arg) CGG ⎭	U C A G
	A	AUU ⎫ AUC ⎬ Isoleucine AUA ⎪ (Ile) AUG Met or start	ACU ⎫ ACC ⎬ Threonine ACA ⎪ (Thr) ACG ⎭	AAU ⎫ Asparagine AAC ⎭ (Asn) AAA ⎫ Lysine AAG ⎭ (Lys)	AGU ⎫ Serine AGC ⎭ (Ser) AGA ⎫ Arginine AGG ⎭ (Arg)	U C A G
	G	GUU ⎫ GUC ⎬ Valine GUA ⎪ (Val) GUG ⎭	GCU ⎫ GCC ⎬ Alanine GCA ⎪ (Ala) GCG ⎭	GAU ⎫ Aspartic GAC ⎭ acid(Asp) GAA ⎫ Glutamic GAG ⎭ acid (Glu)	GGU ⎫ GGC ⎬ Glycine GGA ⎪ (Gly) GGG ⎭	U C A G

Third base

Conceptual Integrated Science — Third Edition

Chapter 16: Genetics
Chromosomes

1. On the left are some chromosomes from a diploid cell. On the right are chromosomes from a human cell.

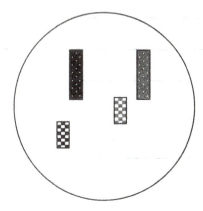

a. How many chromosomes are in the diploid cell? _____

b. After the diploid cell goes through meiosis, how many chromosomes will there be in the
    resulting cells? _____

c. How many chromosomes are in the human cell? _____

d. Two kinds of cells that are produced through meiosis in humans are
    _____ and _____.

e. The human chromosomes above belong to one of the two people shown here. Circle the
    correct person.

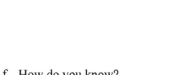

f. How do you know?

    _____

    _____

Conceptual Integrated Science — Third Edition

Chapter 16: Genetics
Comparing Mitosis and Meiosis

1. Give some examples of when organisms use mitosis.

   _____

   _____

   _____

2. a. When do organisms use meiosis?

   _____

   b. What kinds of cells are produced through meiosis?

   _____

Circle the correct answers.

3. a. In meiosis, the number of cells produced is   (one)   (two)   (three)   (four).

   In mitosis, the number of cells produced is   (one)   (two)   (three)   (four).

   b. During meiosis, cells produced are   (haploid)   (diploid).

   During mitosis, cells produced are   (haploid)   (diploid).

   c. The cells produced are   (different from one another)   (identical) in meiosis.

   The cells produced are   (different from one another)   (identical) in mitosis.

   d. Crossing over happens during   (mitosis)   (meiosis).

Conceptual Integrated Science
Third Edition

Chapter 16: Genetics
Dominant and Recessive Traits

1. Some small woodland creatures have either spots or stripes. Fur pattern is determined by a single gene. The striped phenotype is **dominant** and the spotted phenotype is **recessive**.

   Both the woodland creatures shown below are **homozygotes**.

   Genotype is   (aa)   (AA)   (aa)   (AA).

   Phenotype is   (spotted)   (striped)   (spotted)   (striped).

2. The woodland creature below is a **heterozygote**. Does it have spots or stripes? Sketch the woodland creature's fur pattern and circle the genotype and phenotype.

   Genotype is   (aa)   (AA)   (Aa).

   Phenotype is   (spotted)   (striped).

Fill in the blanks.

3. Here are two more woodland creatures. Which of the following must be a **homozygote**? Which could be either a **homozygote** or a **heterozygote**?

   _____   _____

   So we see that the spotted one must be a   (homozygote)   (heterozygote),

   since you have to have two recessive alleles to have spots. The striped one is

   (a homozygote)   (a heterozygote)   (either a homozygote or heterozygote).

┌─ **Conceptual Integrated Science** **Third Edition**

Chapter 16: Genetics
Inheritance Patterns

1. Suppose you breed two woodland creatures together. One individual has genotype aa, and the other has genotype AA. Draw the cross below.

Genotype is   (aa)   (AA)   (aa)   (AA).

Phenotype is   (spotted) (striped)   (spotted) (striped).

2. What are the progeny like?

Genotype is   (aa) (AA) (Aa) and phenotype is (spotted) (striped).

Fill in the table below.

3. Now you breed two of the progeny from Question 2 together. What will their offspring look like?

So, the offspring include woodland creatures that are

(spotted)   (striped)   (both spotted and striped).

The offspring phenotypes are found in a ratio of _____ striped: _____ spotted.

Name _____ Date _____

Conceptual Integrated Science
Third Edition

Chapter 16: Genetics
Genetic Engineering

1. What is genetic engineering?

   _____

   _____

2. Carla learns that some people have a condition in which their bodies do not make enough human growth hormone. Because human growth hormone promotes growth, making too little of the hormone results in unusually short stature. Doctors give these patients injections of human growth hormone in order to help them achieve normal heights.

   Carla asks, "But how do the doctors get human growth hormone for the injections? They can't take hormone from other people, can they?"

   You say, "No, they don't collect it from other people. They use genetically engineered organisms that produce human growth hormone." Explain to Carla how this works.

   _____

   _____

   _____

3. Dirk wants to know why some people are against using genetically engineered organisms in food and agriculture. Explain at least two potential issues.

   _____

   _____

   _____

   _____

   _____

   _____

—Conceptual Integrated Science— Third Edition

Chapter 17: The Evolution of Life
Genetic Variation in Human Traits

1. Do the human traits listed below show genetic variation?

    a. age   _____

    b. eye color   _____

    c. number of toes   _____

    d. curliness or straightness of hair   _____

    e. presence or absence of dimples   _____

    f. upright posture   _____

    g. owning versus not owning a dog   _____

    h. height   _____

2. What kinds of traits have the potential of evolving via natural selection?

_____

Conceptual Integrated Science
Third Edition

Chapter 17: The Evolution of Life
Natural Selection

1. On a tropical island, there lives a species of bird called the Sneezlee bird. Sneezlee birds eat seeds and also show genetic variation in beak size. Some Sneezlee birds have bigger beaks, and some have smaller beaks.

   a. Draw a box around the bird with the biggest beak.

   b. Put a check mark by the bird with the smallest beak.

2. One year, there are not many seeds available for Sneezlee birds due to a drought in the summer. Small seeds are quickly eaten by the Sneezlee birds. Only larger, tougher seeds are left. Sneezlee birds with larger, stronger beaks are better at cracking these larger seeds.

   a. Draw a box around the two birds most likely to survive the drought.

   b. Mark Xs through two Sneezlee birds that are more likely to die.

3. Beak size is an inherited trait. Parent birds with larger, stronger beaks tend to have offspring with larger, stronger beaks. How would the Sneezlee bird population evolve due to the drought?

   _____

   _____

Chapter 17: The Evolution of Life
Size and Shape

1. The imaginary mammal below occupies temperate forests in the Eastern United States.

a. If a population of these mammals moved and successfully colonized an Arctic habitat,
   how might you predict that it would evolve? Draw the Arctic form of the mammals below.

> Think about ear and leg length

b. If a population of these mammals moved and successfully colonized a desert habitat,
   how might you predict that it would evolve? Draw the desert form of the mammals below.

> Think area vs volume.

c. Explain your drawings.

_____

_____

_____

_____

_____

_____

_____

_____

_____

Chapter 17: The Evolution of Life
Mechanisms of Evolution

Suppose you have a population of peppered moths in which the allele frequencies are 50% light allele and 50% dark allele. For each of the events below, list the mechanism of evolution involved and the event's effect on allele frequencies.

Event	Mechanism of evolution (natural selection, mutation pressure, genetic drift, or gene flow)	Effect on allele frequencies (increases frequency of the light allele, increases frequency of the dark allele, or change cannot be predicted)
1. Some light moths migrate into the population from a nearby unpolluted area.		
2. Pollution in the town increases.		
3. A storm kills half the moths in the population.		
4. Some dark moths migrate into the population from a nearby polluted area.		
5. Just by chance, dark moths leave more offspring than light moths one year.		
6. Dark moths survive better than light moths.		

Conceptual Integrated Science
Third Edition

Chapter 17: The Evolution of Life
Speciation

1. Define the following terms:

   Species _____

   Speciation _____

   Reproductive barrier _____

2. Explain the difference between a prezygotic reproductive barrier and a postzygotic reproductive barrier. Then, give one example of each.

   _____

   _____

   _____

   _____

   _____

3. A geographic barrier–a newly formed river–divides a population of frogs into two separate populations. Explain how this event could result in speciation.

   _____

   _____

   _____

   _____

Conceptual Integrated Science
Third Edition

Chapter 18: Diversity of Life on Earth
Classification

Circle the correct answer:

1. Linnaean classification involves grouping species together on the basis of how

    (similar)   (different) they are.

2. Fill in the levels of Linnaean classification from the largest group to the smallest group.

    <u>Domain</u>

    <u>Kingdom</u>

    _____

    _____

    _____

    _____

    _____

    _____

Circle the correct answers:

3. The scientific name of a species consists of its (genus)   (family)   (species) name

    and its (genus)   (family)   (species) name.

Earth is home to as many as 10 million different species! Humans are just one of these.

Name _____ Date _____

Chapter 18: Diversity of Life on Earth
Evolutionary Trees

Circle the correct answers:

1. Biologists now try to classify organisms on the basis of how (similar) (closely related) they are to each other.

2. The evolutionary tree below shows how three species—the daisy, the honey mushroom, and the gentoo penguin—are related.

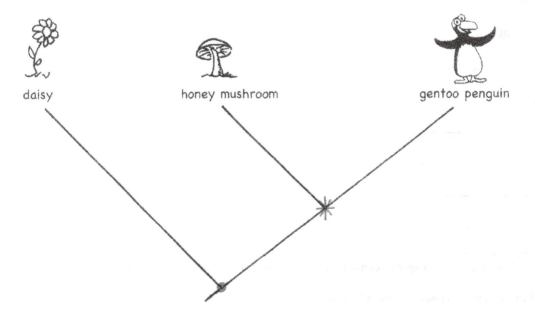

daisy   honey mushroom   gentoo penguin

    a. This evolutionary tree tells us that the (daisy) (honey mushroom) (gentoo penguin)

       and (daisy) (honey mushroom) (gentoo penguin) are more closely related to each

       other than either is to the (daisy) (honey mushroom) (gentoo penguin).

    b. Place a dot at the point where the lineage that eventually gave rise to daisies split from the lineage that eventually gave rise to gentoo penguins.

    c. Now place an asterisk at the point where the lineage that eventually gave rise to honey mushrooms split from the lineage that eventually gave rise to gentoo penguins.

— Conceptual Integrated Science — Third Edition

Chapter 18: Diversity of Life on Earth
Bacteria and Archaea

Fill in the blanks:
1. The three domains of life are

   _____, _____, and _____.

   Of the 3 domains, _____ and _____ consist of

   prokaryotes and _____ consists of eukaryotes.

Circle the correct answers:
2. a. Are there any bacteria that can make their own food through photosynthesis, the way plants do? (Yes)   (No)

   b. Are there any bacteria that get their food from other organisms, the way animals do? (Yes)   (No)

3. Bacteria reproduce (sexually)   (asexually).

4. Bacteria that live on our bodies benefit us by

   (producing vitamins)

   (keeping dangerous bacteria from invading our bodies)

   (both of these).

5. Are all archaea extremophiles? (Yes)   (No)

6. Some chemoautotrophs make food using

   (chemical energy)

   (energy from sunlight).

In life, it's the stuff you can't see that does us the most damage.

Conceptual Integrated Science
Third Edition

Chapter 18: Diversity of Life on Earth
Domain Eukarya and Protists

1. All the living things found in the domain Eukarya have   (prokaryotic cells)   (eukaryotic cells).

2. Name the four kingdoms that make up the domain Eukarya.

   _____

   _____

   _____

   _____

3. Protists get food from photosynthesis or from other organisms. Place the protists below into the correct column.

   Amoebas

   Kelp

   Diatoms

   Ciliates

   Plasmodium, the protist that causes malaria

Kelp is a protist!

   Get food from photosynthesis   Get food from other organisms

   _____   _____

   _____   _____

   _____   _____

   _____   _____

Name _____ Date _____

Conceptual Integrated Science
Third Edition

Chapter 18: Diversity of Life on Earth
Plants

1. Match the following plant structures with their function.

   Roots   _____   a. Distributes water and other resources

   Shoots   _____   b. Conduct photosynthesis

   Vascular system   _____   c. absorb water and nutrients from soil

Circle the correct answers:

2. Mosses (do)   (do not) have vascular systems whereas ferns

   (do)   (do not) have vascular systems.

3. (Mosses)   (Ferns)   (Mosses and ferns)   (Seed plants) have to live in a moist environment because sperm must swim through the environment to fertilize eggs.

4. Describe each structure and write the name of the group of plants that possess each of the structures.

Structure	Description	Plant group in which structure is found
Pollen	_____	_____
	_____	
Seed	_____	_____
	_____	
Flower	_____	_____
	_____	
Fruit	_____	_____
	_____	
Cone	_____	_____

┌─ **Conceptual Integrated Science** ─ Third Edition

Chapter 18: Diversity of Life on Earth
Animals

Match the following animal groups with the list of features on the right. Note that some groups have more than one answer!

Sponges	_____
Cnidarians	_____
Flatworms	_____
Roundworms	_____
Arthropods	_____
Mollusks	_____
Annelids	_____
Echinoderms	_____
Chordates	_____
Cartilaginous fishes	_____
Ray-finned fishes	_____
Amphibians	_____
Reptiles	_____
Birds	_____
Mammals	_____

a. Clams, oysters, and squids are all part of this group.
b. The swim bladder of these animals makes their density the same as the density of water—this makes it much easier for them to swim well!
c. These animals sink if they stop swimming.
d. Many of the animals in this group start life in the water and then move to land as adults.
e. All their muscles run from head to tail. Because of this, these animals move like flailing whips.
f. There is a sedentary polyp stage and a swimming medusa stage.
g. Leeches belong to this group.
h. This group includes worms whose bodies are divided into segments.
i. Their bodies are divided into segments and their legs have bendable joints.
j. Their skins are made of living cells that can dry out, so they have to stay in moist environments.
k. In these animals, a constant flow of water comes in through many pores and goes out the top. The purpose of this constant flow is to catch food.
l. Birds and crocodiles are examples of this group of animals.
m. These animals move using tube feet.
n. Watch out! Their tentacles have barbed stinging cells.
o. They have wings and hollow bones.
p. The insects are part of this group.
q. The vertebrates are part of this group.
r. Starfish belong to this group.
s. These animals shed their exoskeleton as they grow.
t. These animals, such as sharks and rays, have a skeleton made of cartilage
u. These animals have hair and feed their young milk.
v. Tapeworms -parasites that live in the intestines of humans and other animals- belong to this group.
w. The platypus belongs to this group, and so do bats and humans!
x. Most of the animals in this group have a shell, although slugs and octopuses don't.
y. They are flying endotherms.
z. Frogs belong to this group.

Conceptual Integrated Science — Third Edition

Chapter 18: Diversity of Life on Earth
Viruses and Prions

1. Is a virus a prokaryote, a eukaryote, or neither? Explain your answer.

    _____

    _____

2. Describe the structure of a virus.

    _____

    _____

    _____

3. How do viruses "reproduce"?

    _____

    _____

    _____

    _____

4. What is a prion? How does a prion cause disease?

    _____

    _____

    _____

─Conceptual Integrated Science ── Third Edition

Chapter 19: Human Biology I—Control and Development

Parts of the Brain

Match the parts of the brain with their body functions.
Note that some parts have more than one function.

Brainstem	_____	a. Deals with visual information (what we see)
Cerebellum	_____	b. Controls balance, posture, and coordination
Cerebrum	_____	c. Deals with sensory information about temperature, touch, and pain
Thalamus	_____	d. Controls basic involuntary activities such as heartbeat, respiration, and digestion
Hypothalamus	_____	e. Allows us to understand spoken language

f. Controls our voluntary movements

g. Sorts and filters information and then passes it to the cerebrum

h. Responsible for emotions such as pleasure and rage

i. Controls our speech

j. Controls hunger, thirst, and sex drive

k. Controls the fine movements we use in activities that we perform "without thinking"

Conceptual Integrated Science — Third Edition

Chapter 19: Human Biology I—Control and Development
The Nervous System

1. The two main parts of the nervous system are the _____

   and _____.

   The central nervous system consists of the _____ and the

   _____.

2. The three types of neurons are _____,

   _____, and _____.

   Messages from the senses to the central nervous are carried by

   _____. Neurons that connect one neuron to another neuron are

   _____. Messages are carried from the central nervous system

   to muscle cells or to other responsive organs by _____.

3. Motor neurons are further divided into two groups:

   the _____,

   which controls voluntary actions and stimulates our voluntary muscles,

   and the _____,

   which controls involuntary actions and stimulates our involuntary muscles

   and other internal organs.

   The autonomic nervous system includes a _____

   that promotes a "fight or flight" response and a _____

   that operates in times of relaxation.

Conceptual Integrated Science — Third Edition

Chapter 19: Human Biology I—Control and Development
Parts of a Neuron

1. a. Label the parts of the neuron in the diagram.

   Dendrites

   Cell body

   Axon

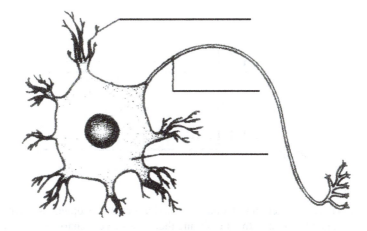

   b. Explain the function of each part of a neuron.

Part of a neuron	Function
Dendrites	_____
Cell body	_____
Axon	_____

Lots to learn ...
lots to know.

Conceptual Integrated Science
Third Edition

Chapter 19: Human Biology I—Control and Development
Action Potentials

1. This is a neuron at rest. The neuron is at its resting potential. Draw a + sign on the side of the membrane that is positively charged. Draw a – sign on the side of the membrane that is negatively charged.

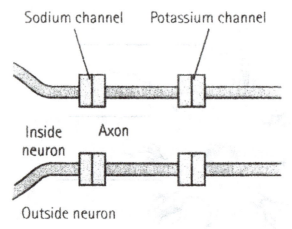

2. Now the neuron fires! There is an action potential. The sodium channels open. Use arrows to show how the sodium ions move. Draw a + sign on the side of the membrane that is positively charged. Draw a – sign on the side of the membrane that is negatively charged.

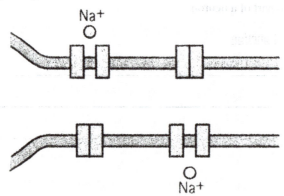

3. Now the sodium channels close, and the potassium channels open. Use arrows to show how the potassium ions move. Draw a + sign on the side of the membrane that is positively charged. Draw a – sign on the side of the membrane that vis negatively charged. The action potential is over.

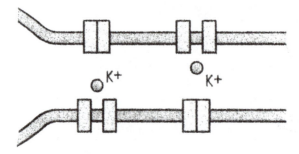

Conceptual Integrated Science
Third Edition

Chapter 19: Human Biology I—Control and Development
Senses

1. The light-sensitive cells are found in the part of the eye

   called the _____.

   The two types of light-sensitive cells are _____

   and _____.

2. State whether the following describe rods or cones.

   a. Vision at night or in dim light   _____

   b. Let us see color   _____

   c. Detect only black, white, and shades of gray   _____

   d. Not very good at making out fine details   _____

   e. Nonfunctioning version of these causes colorblindness   _____

   Number from 1 to 4:

3. The ear consists of 3 parts: the outer, middle, and inner ear. Sound moves through the air into the ear
   in the following order:

   _____ middle ear bones

   _____ pinna

   _____ cochlea

   _____ eardrum

4. List the five basic tastes.

   _____ _____ _____

   _____ _____

Conceptual Integrated Science
Third Edition

Chapter 19: Human Biology I—Control and Development
Hormones

1. What is a hormone?

   _____

   _____

2. What are the two types of hormones? List 2 differences between them.

   _____

   _____

   _____

   _____

   _____

3. Explain the function of each of the following hormones.

   Insulin   _____

   Melatonin   _____

   Antidiuretic hormone   _____

   Epinephrine   _____

   Parathyroid hormone   _____

   Oxytocin   _____

4. Explain how hormones are involved in maintaining homeostasis in the body.

   _____

   _____

   _____

   _____

Conceptual Integrated Science
Third Edition

Chapter 19: Human Biology I—Control and Development
Skeleton

1. List three functions of the skeleton.

   _____

   _____

   _____

2. Label the three layers of bones.

   Bone marrow

   Compact bone

   Spongy bone

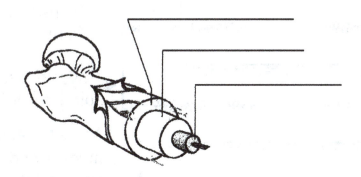

3. What is the function of red bone marrow?

   _____

   _____

No bones about it!

Conceptual Integrated Science Third Edition

Chapter 19: Human Biology I—Control and Development
Muscle Contraction

1. Muscles work by   (lengthening)   (shortening).

2. Two kinds of proteins are involved in muscle contraction:

   _____ and

   _____.

3. Draw the myosin heads (ovals) during muscle contraction in the diagram below.

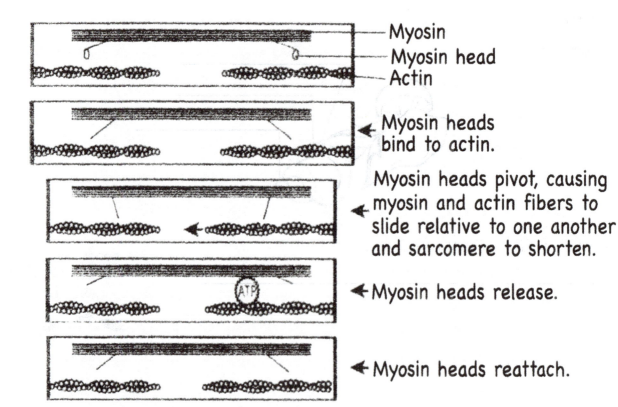

Myosin
Myosin head
Actin

◄ Myosin heads
  bind to actin.

◄ Myosin heads pivot, causing
  myosin and actin fibers to
  slide relative to one another
  and sarcomere to shorten.

◄ Myosin heads release.

◄ Myosin heads reattach.

◄ Myosin heads pivot again,
  causing further muscle
  contraction.

Conceptual Integrated Science
Third Edition

Chapter 20: Human Biology II—Care and Maintenance
Integration of Body Systems

1. The circulatory system and respiratory system work together to help keep the body's tissues supplied with oxygen. Explain how each of the two systems contributes.

   _____

   _____

   _____

2. Which body systems are involved in the task you are performing now, reading this question and writing your response?

   _____

   _____

   _____

   _____

   _____

3. Choose another activity that you perform regularly—such as eating a meal, going swimming, brushing your hair, or another activity of your choice—and explain how multiple body systems contribute to that activity.

   _____

   _____

   _____

   _____

   _____

Name_____ Date _____

Conceptual Integrated Science
Third Edition

Chapter 20: Human Biology II—Care and Maintenance
Circulatory System

Fill in the blanks:

1. Each heartbeat begins in a part of the right atrium called the _____, or pacemaker.

   The pacemaker starts an action potential that sweeps quickly through two chambers of the

   heart, the _____ and _____,

   which contract simultaneously.

   The signal also travels to the _____,

   and from there to the other two chambers of the heart,

   the _____

   and _____.

   These two chambers also contract simultaneously.

2. a. The sound of a heartbeat is "lub-dubb, lub-dubb." What is the "lub"?

   _____

   _____

   b. What is the "dubb"?

   _____

   _____

3. The three types of cells found in blood are _____,

   _____, and _____.

   _____ transport oxygen to the body's tissues.

   _____ are part of the immune system and help our bodies

   defend against disease. _____ are involved in blood clotting.

4. The molecule in red blood cells that carries oxygen is called _____.

Conceptual Integrated Science Third Edition

Chapter 20: Human Biology II—Care and Maintenance

Circulatory System—continued

Number from 1 to 9:
5. In what order does blood flow around the body? Begin with the right atrium.

_____   Arteries to lungs

_____   Right atrium

_____   Left atrium

_____   Veins from body tissues

_____   Veins from lungs

_____   Capillaries near body tissues

_____   Arteries to body tissues

_____   Right ventricle

_____   Left ventricle

Conceptual Integrated Science
Third Edition

Chapter 20: Human Biology II—Care and Maintenance
Respiratory System

1. Match each part of the respiratory system with its description.

Nasal passages	_____	a. Where gas exchange occurs
Larynx	_____	b. Another word for trachea
Trachea	_____	c. Structure allows us to speak
Alveoli	_____	d. Raises the ribcage when we inhale
Diaphragm	_____	e. Smelling happens here
Rib muscles	_____	f. Dome-shaped muscle helps us inhale
Bronchi	_____	g. Tubes that go to the right and left lungs
Windpipe	_____	h. A short tube stiffened by rings of cartilage

2. Which figure shows a person inhaling? Which figure shows a person exhaling?

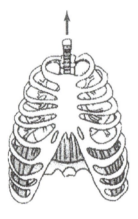

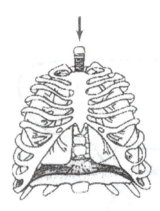

a. _____

b. _____

Conceptual Integrated Science
Third Edition

Chapter 20: Human Biology II—Care and Maintenance
Digestion

1. Why do we have to digest our food?

_____

_____

2. Label the parts of the digestive system in the diagram using the following terms:

Stomach

Liver

Pancreas

Small intestine

Esophagus

Large intestine

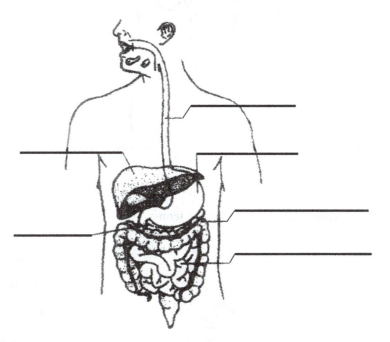

3. Where does each of the following events important in digestion occur?

a. _____   Bile is made

b. _____   A highly acidic mix of hydrochloric acid and
                                          digestive enzymes is added

c. _____   Most nutrients are absorbed into the body

d. _____   Food is chewed and down into smaller pieces

e. _____   Water is absorbed

f. _____   Muscular churning of food

g. _____   Saliva begins digesting starches in our food

h. _____   Enzymes from the pancreas help with digestion

i. _____   Vitamins K and B are made by bacteria

Conceptual Integrated Science
Third Edition

Chapter 20: Human Biology II—Care and Maintenance
Excretory System

1. Label the parts of the excretory system in the diagram using the following terms:

Urethra

Bladder

Ureter

Kidney

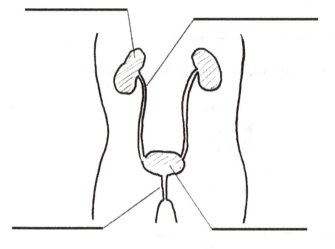

2. The functional unit of the kidney is called the _____.

3. Label the structures of a nephron using the following terms:

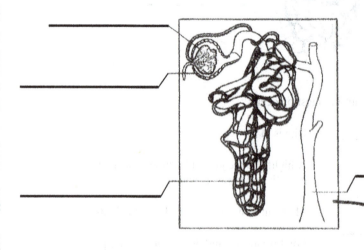

Capillaries

Bowman's capsule

Loop of Henle

Collecting duct

Nephron
Renal pelvis

Ureter

Conceptual Integrated Science — Third Edition

Chapter 20: Human Biology II—Care and Maintenance

Excretory System—continued

4. Fluid leaves the _____ of the circulatory system and enters the

part of the nephron called _____.

5. The two parts of the nephron where water is reabsorbed by the body are the

_____ and the _____.

The attention a community gives to the maintenance of its infrastructure is a measure of the value of that community. Likewise with the care and maintenance of your body.

Conceptual Integrated Science — Third Edition

Chapter 20: Human Biology II—Care and Maintenance
Immune System

1. Describe the role of each part of the immune system.

   a. _____

   b. Mucus _____

   _____

   c. Enzymes in tears and milk _____

   d. Inflammatory response _____

   _____

   _____

   e. Antibodies _____

   _____

   _____

   f. Clones _____

   _____

   g. B cells _____

   h. T cells _____

   i. Memory cells _____

   _____

   _____

2. Explain how a vaccine works. _____

_____

_____

_____

_____

_____

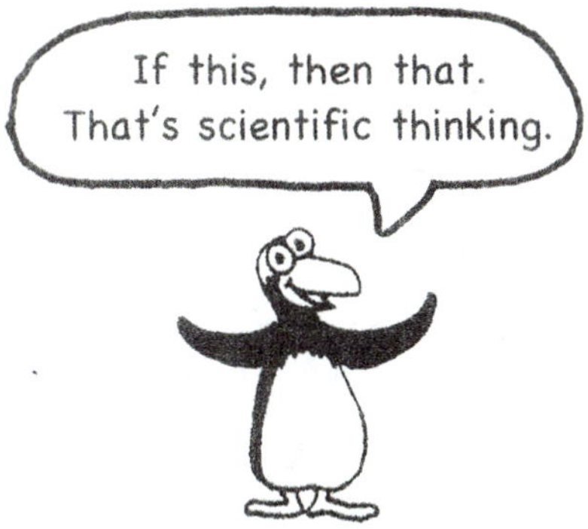
If this, then that.
That's scientific thinking.

Name _____ Date _____

Conceptual Integrated Science

Chapter 21: Ecology
Populations, Communities, and Ecosystems

1. Define the following terms.

   Ecology   _____

   Population   _____

   Community   _____

   Ecosystem   _____

                     _____

2. In the space below, draw three (or more) populations of organisms in their habitat. Then, add labels to your drawing to identify:

   a. each of the three populations

   b. at least one way in which two of the populations interact with each other

   c. at least one way in which one population interacts with the abiotic features of its environment

   (For example, you might draw plants, butterflies, and birds in your backyard. You could then describe an interaction between the butterflies and plants in which butterflies feed on nectar from the plant's flowers. Finally, you could explain how the birds drink water from a fountain in the yard. Come up with your own example!)

Conceptual Integrated Science
Third Edition

Chapter 21: Ecology
Exponential Growth and Logistic Growth

1. Which of the following graphs shows exponential growth and which shows logistic growth?

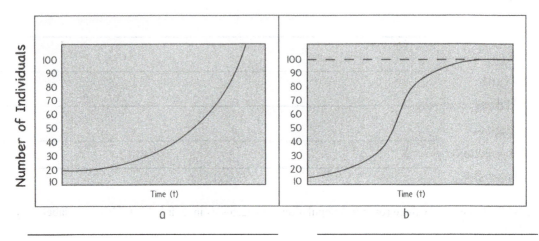

a _____   b _____

Circle the correct answers:

2. The carrying capacity in the graph on the right is   (10)   (100)   (more than 100) individuals.

   Appendix D will help you with Questions 3 and 4.

3. Exponential growth is nicely illustrated with the children's story of a rapidly growing beanstalk that doubles in height each day.

   Suppose one day after breaking ground the stalk is 1 centimeter high.

   If growth is continual, at the end of the second day it will be   (1)   (2)   (4) cm high.

   At the end of the third day it will be   (1)   (2)   (4) cm high.

   Doubling each day results in exponential growth so that on the 36th day it reaches the Moon! Working backward, the height of the beanstalk on the 35th day was (one-half)   (one-quarter)   (one-third) the distance from Earth to the Moon.

   And on the 34th day the beanstalk was (one-half)   (one-quarter)   (one-third) the distance from Earth to the Moon.

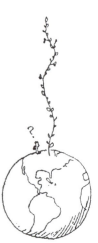

4. Then there is the story of a lily pond with a single leaf. Each day the number of leaves doubles; on the 30th day the pond is completely full.

   On what day was the pond half covered? (15 days)   (28 days)   (29 days)

   On what day was it one-quarter covered? (15 days)   (28 days)   (29 days)

Conceptual Integrated Science — Third Edition

Chapter 21: Ecology

Species Interactions

Circle the correct answers:
Look at the food chain below.

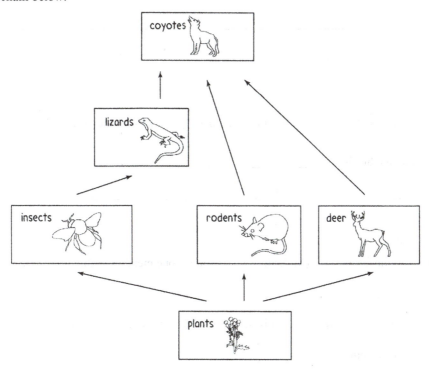

a. The producers in this community are (insects)   (plants)   (rodents).

b. The primary consumers in this community are

   (plants, insects, and rodents)   (insects, rodents, and deers)   (insects, lizards, and coyotes).

c. The secondary consumers in this community are

   (plants, insects, and rodents)   (insects, rodents, and deers)   (lizards and coyotes).

d. The tertiary consumers in this community are (deer)   (coyotes)   (insects).

e. The top predators in this community are (deer)   (coyotes)   (insects).

┌─ **Conceptual Integrated Science** ─── **Third Edition**

Chapter 21: Ecology
Species Interactions—continued

3. Briefly describe a species' niche.

_____

_____

Fill in the blanks:

4. In a symbiosis, individuals of two species live in close association with one another.

a. The three types of symbiosis are _____,

_____, and _____.

Circle the correct answers:

b. (Parasitism)   (Commensalism)   (Mutualism) is good for one member of the interaction

and bad for the other.

c. (Parasitism)   (Commensalism)   (Mutualism) is good for one species in the interaction

and has no effect on the other.

d. (Parasitism)   (Commensalism)   (Mutualism) is a relationship that benefits both species

involved.

Conceptual Integrated Science — Third Edition

Chapter 21: Ecology
Biomes and Aquatic Habitats

1. Match each of the following features with the appropriate biome:

   a. Tropical grassland   _____   Tropical forest

   b. A habitat that receives very little precipitation,   _____   Temperate forest
      can be cold or hot
                                                              _____   Coniferous forest
   c. Permafrost is found in this biome
                                                              _____   Tundra
   d. Mild, rainy winters, and hot summers with
      drought and fire   _____   Savanna

   e. The trees in this biome have needlelike   _____   Temperate grassland
      leaves that can survive cold winters
                                                              _____   Desert
   f. More species are found in this biome than
      in all other biomes combined   _____   Chaparral

   g. A grassland found in areas with four distinct seasons

   h. In this biome, trees drop their leaves in the autumn

2. What kinds of adaptations do you expect to see in a freshwater organism that lives in a river
   or stream? Why?

   _____

   _____

   _____

3. a. What is an estuary?

   _____

   b. What adaptations are found in plants that live in estuaries?

   _____

   _____

Conceptual Integrated Science Third Edition

Chapter 21: Ecology
Biomes and Aquatic Habitats—continued

4. a. What is an intertidal habitat?

   _____

   b. What feature of intertidal habitats makes them challenging environments to live in?

   _____

   _____

   _____

   _____

   c. What are some adaptations found in organisms that live in intertidal habitats?

   _____

   _____

   _____

   _____

Conceptual Integrated Science
Third Edition

Chapter 21: Ecology
Carbon Cycle and Nitrogen Cycle

1. What is a biogeochemical cycle? Why does the term include both "bio" and "geo"?

   _____

   _____

   _____

2. Draw one possible path of a carbon atom from the atmosphere, to a plant, to an animal, and back to the atmosphere. In your drawing, indicate where "photosynthesis" occurs and where "cellular respiration" occurs.

3. How does nitrogen move from the abiotic world to the biotic world?

   _____

   _____

4. How does nitrogen move from the biotic world to the abiotic world?

   _____

Conceptual Integrated Science — Third Edition

Chapter 21: Ecology
Ecological Succession

1. What is ecological succession?

   _____

   _____

2. What are some differences between primary succession and secondary succession?

   _____

   _____

   _____

3. What is a pioneer species?

   _____

   _____

   _____

4. What is a climax community?

   _____

   _____

5. Hoes does biodiversity change during ecological succession?

   _____

6. Explain the intermediate disturbance hypothesis.

   _____

   _____

   _____

   _____

Conceptual Integrated Science
Third Edition

Chapter 22: Plate Tecton: The Earth System
How Many Layers?

The three-layer model

When you were younger, you may have learned that Earth has *three* layers: crust, mantle, and core. This simple model is based on Earth's *composition*, or chemical makeup. But Earth scientists (and science students like you) know that there is another way to look at our planet. Scientists usually divide Earth into five structural layers on the basis of physical properties: *lithosphere, asthenosphere, mesosphere, outer core,* and *inner core.*

1. a. Label the three layers:

   Crust

   Mantle

   Core

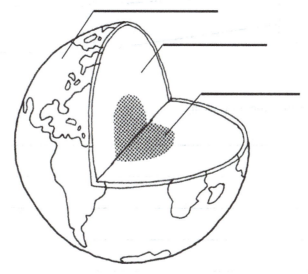

Fill in the blanks:

   b. In this three-layer model, Earth's surface layer is called the _____.

   It consists mostly of low-density rocks such as _____ and

   _____.

   Earth's middle layer is called the _____. It consists of

   _____ that contain dark and dense elements such as magnesium.

   Earth's innermost layer is mostly made of _____.

Conceptual Integrated Science
Third Edition

Chapter 22: Plate Tecton: The Earth System
How Many Layers?—continued

The five-layer model

1. a. Label the five layers:

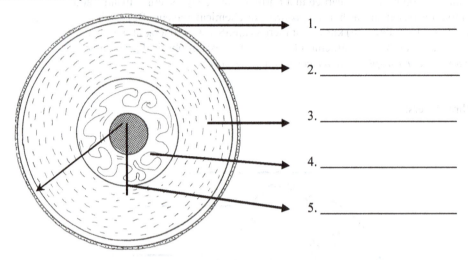

1. _____

2. _____

3. _____

4. _____

5. _____

b. Describe the physical properties of each layer.

_____

_____

_____

_____

_____

_____

Circle True or False:

2. a. The asthenosphere is a soft layer of the mantle on which pieces   (True)   (False)
        of lithosphere move.

    b. The lithosphere is made of two parts—the crust and the upper mantle.   (True)   (False)

    c. Even though the lithosphere is made of different types of rock,   (True)   (False)
        it behaves as a single unit.

    d. The lithosphere is a rigid layer of brittle rock.   (True)   (False)

In the drawing above do the following:

3. a. With a red crayon or colored pencil, color the hottest layer of Earth.

    b. With a blue crayon or colored pencil, color the coolest part of Earth.

    c. Draw an arrow to show the direction of net heat flow within Earth.

Conceptual Integrated Science Third Edition

Chapter 22: Plate Tecton: The Earth System
Get the Picture—Plate Boundaries

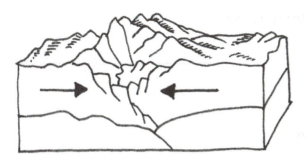

Figure A

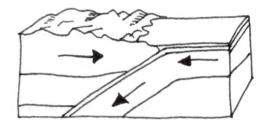

Figure B

Figure C

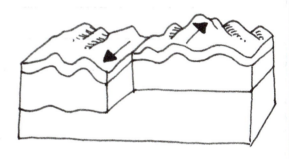

Figure D

Look at the diagrams of plate boundaries.

Answer the following questions:

1. What kind of tectonic plate collision does Figure A show? _____

   _____

2. What kind of tectonic plate collision does Figure B show? _____

   _____

3. What type of plate boundary does Figure C show? _____

4. What type of plate boundary does Figure D show? _____

5. Which diagram shows the creation of new lithosphere? _____

6. Which diagram shows the destruction of old lithosphere? _____

7. Which diagram shows the type of plate collision that creates some of the world's tallest

   boundaries? _____

┌─Conceptual Integrated Science ──── Third Edition

Chapter 22: Plate Tecton: The Earth System
Get the Picture—Plate Boundaries—continued

8. What geological events occur most often near plate boundaries?

_____

_____

9. Why are plate boundaries active geological regions?

_____

_____

_____

_____

Physics reaches to chemistry.
Both physics and chemistry reach to biology.
All three reach to Earth science!

10. a. *Challenge*: Do plate boundaries last forever? _____

   b. Explain.

_____

_____

_____

_____

_____

The answer to number 10 isn't in the textbook, and requires *critical thinking*. Good Luck!

Conceptual Integrated Science Third Edition

Chapter 22: Plate Tecton: The Earth System
Get the Picture—Seafloor Spreading

1. The diagram shows what happens during seafloor spreading. But it isn't very useful unless you can label all the parts. Match the numbers (1 to 5) to the terms below.

_____   convection

_____   magma rising

_____   mid-ocean ridge

_____   older crust

_____   younger crust

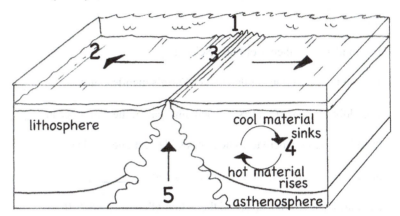

2. During seafloor spreading, magma rises to fill a gap between diverging tectonic plates.
   After magma erupts it is called *lava*.
   Draw in the lava that would erupt at this spreading center.

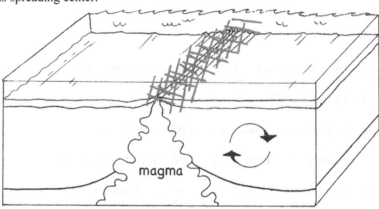

3. What happens to lava after it cools on the seafloor?

_____

4. Why is the process in the diagram called *seafloor spreading*?

_____

_____

_____

Conceptual Integrated Science — Third Edition

Chapter 22: Plate Tecton: The Earth System
Play Tectonics

Use the clues below to fill in the words of the crossword puzzle in the numbered blanks on the facing page.

Across

1. The fundamental theory of Earth Science

5. The mode of heat transfer that makes Earth's mantle churn

7. The global system of undersea mountains, called the "mid-ocean _____."

8. Plate boundaries where tectonic plates slide past one another

9. Inside ocean trenches, the temperature is very _____.

10. German scientist who advanced the notion of continental drift

12. The process by which gravity contributes to plate motion

14. The recycling of lithosphere at an ocean trench

Down

2. Earth's rigid layer, which is broken up into tectonic plates

3. What gets wider when new seafloor is created?

4. Earth's plastic layer, which tectonic plates slide over

6. What escapes from Earth's interior and drives the motion of tectonic plates?

11. The branch of Earth Science concerned with Earth's interior

13. Another name for chunks of lithosphere

15. Old lithosphere sinks at ocean trenches because it is cooler and therefore more _____ than the young lithosphere forming near spreading centers.

16. A deep crack in the ocean floor where old lithosphere is subducted

Chapter 22: Plate Tecton: The Earth System

Play Tectonics—continued

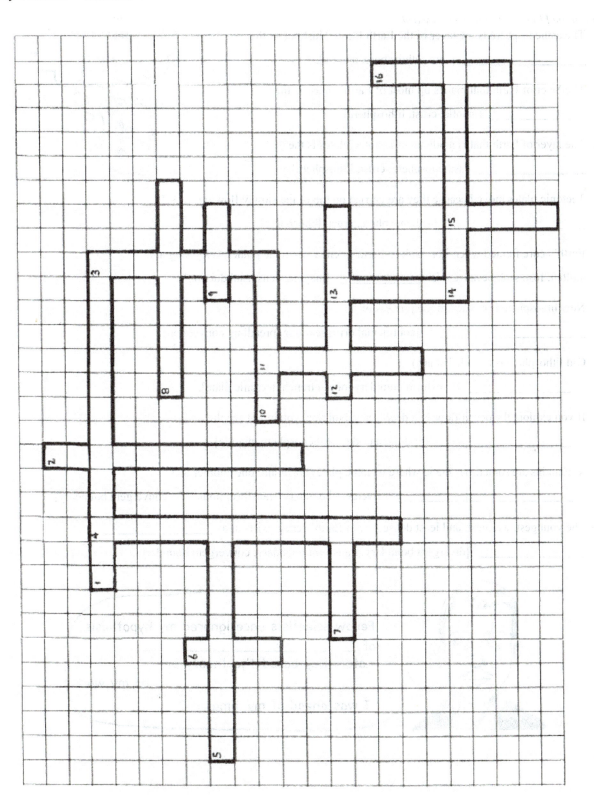

Conceptual Integrated Science
Third Edition

Chapter 22: Plate Tecton: The Earth System
Plate Tectonics Vocabulary Review

Fill in the blanks with the correct word:

> lithosphere   crust
> tectonic plates
> mantle

1. The asthenosphere is the layer of the Earth located just below the

   _____ [core, lithosphere, mantle].

2. The layer of Earth that makes up most of Earth's mass is the

   _____ [mantle, crust, lithosphere].

3. The layer of Earth that is made up of tectonic plates is the

   _____ [asthenosphere, crust, lithosphere].

4. Tectonic plates move because they are carried on top of the slowly flowing

   _____ [asthenosphere, crust, lithosphere].

5. Earth's core formed when iron sank to Earth's center when the planet was young and

   molten. Iron sank because it is _____ [magma, dense, subducting].

6. New lithosphere is created in the process of

   _____ [continental drift, seafloor spreading, convection].

7. Old lithosphere is recycled at a(n)

   _____ [transform boundary, ocean trench, tectonic plate].

8. If you explore the ocean floor in a deep-sea submarine, you might see the

   _____ [mid-ocean ridge, asthenosphere, mantle].

9. Mountains are sometimes made where tectonic plates crash into one another at a

   _____ [divergent boundary, transform boundary, convergent boundary].

10. The youngest, warmest, and least dense part of Earth's crust exists at a _____

    _____ [divergent boundary, transform boundary, convergent boundary].

> Fellow scientists once ignored my hypothesis
> of _____.
> Later they based the theory of
> _____ on my work.
> I was ahead of my time!

Conceptual Integrated Science
Third Edition

Chapter 23: Rocks and Minerals
Mineral Detective

Use the clues given in Sections 23.2 and 23.3 of your textbook to identify each mineral.

1.
Clues:
Silicate
Nonferromagnesian
Makes up about half
   of Earth's crust

Name: _____

2.
Clues:
Nonsilicate
Hardness on Moh's
   scale = 10
Polymorph of graphite

Name: _____

3.
Clues:
Nonsilicate
Evaporite
Cubic crystal form
Tastes good on
   popcorn

Name: _____

4.
Clues:
Nonsilicate
Native element
Gold color
Specific Gravity = 19

Name: _____

5.
Clues:
Contains the silicate
   tetrahedron
Common in Earth's mantle
Is a ferromagnesian silicate
Greenish in color

Name: _____

6.
Clues:
Silicate
Nonferromagnesian
Softer than topaz, harder
   than feldspar
Amethyst is one variety

Name: _____

7.
Clues:
Nonsilicate
Hardness on Moh's
   scale = 3
Perfect cleavage in
   three dimensions

Name: _____

Conceptual Integrated Science
Third Edition

Chapter 23: Rocks and Minerals
Rock Boxes

All rocks can be classified into three basic types, on the basis of how they form: *sedimentary*, *metamorphic*, and *igneous*. There are four cardboard boxes (SEDIMENTARY, METAMORPHIC, IGNEOUS, ALL ROCKS) labeled on the left. Decide which type of rock is being described on the right and write the letter (A to L) that belongs on the dotted line of each box. If a description applies to all three types, write the letter in the **ALL ROCKS** box.

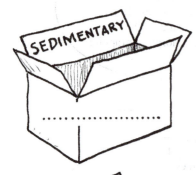

A. Rocks made of magma that has cooled to the solid state

B. Rocks made of bits of preexisting rocks

C. Some of these rocks have large crystals because they slowly cool underground.

D. Granite is an example.

E. Rocks that participate in the rock cycle

F. Sandstone is an example.

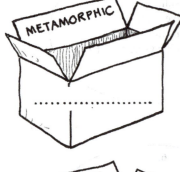

G. Rocks that cover most of Earth's surface

H. Marble is an example.

I. Rocks made when the crystal structure of igneous, sedimentary, or metamorphic rocks is changed

J. Rocks that may be foliated or nonfoliated

K. Rocks that are made of minerals

L. Rocks that may feature fossils

Tinkering is the process of "nosing" your way toward a solution to something you don't quite know how to do—a combination of discovery and play.

┌───

Conceptual Integrated Science ── Third Edition

Chapter 23: Rocks and Minerals
Are You a Rock Hound?

Are rocks your hobby? If you know a lot about rocks, do this crossword puzzle on the facing page and see if you qualify as an official "rock hound."

Use the clues below to fill in the words of the crossword puzzle in the numbered blanks.

Across

1. What determines a mineral's crystal form?

5. Over half of Earth's crust is made of this mineral.

6. What class of minerals includes quartz, feldspar, and others that contain silicon and oxygen?

8. Most minerals form when this cools to the solid state.

9. A rock is a solid _____ of minerals.

11. Rocks that are formed when preexisting rocks recrystallize due to temperature or pressure changes

12. Intrusive igneous rocks (such as granite), form here.

Down

2. Minerals of the same kind have the same crystal structure and _____ composition.

3. Minerals that exhibit this property contain planes of atoms that are weakly bonded to one another.

4. This mineral, the second most common in Earth's crust, is clear when pure but can be yellow, purple, white, pink, or brown when it contains impurities.

7. What chemical element makes up almost half of Earth's crust, yet most people think it is a gas found in air?

10. Rocks formed by magma that cools and solidifies, making mineral crystals that stick together

13. A mineral deposit that is large and pure enough to be mined for a profit

14. A tunnel, pit, or strip of bulldozed earth from which minerals are removed

Chapter 23: Rocks and Minerals

Are You a Rock Hound?—continued

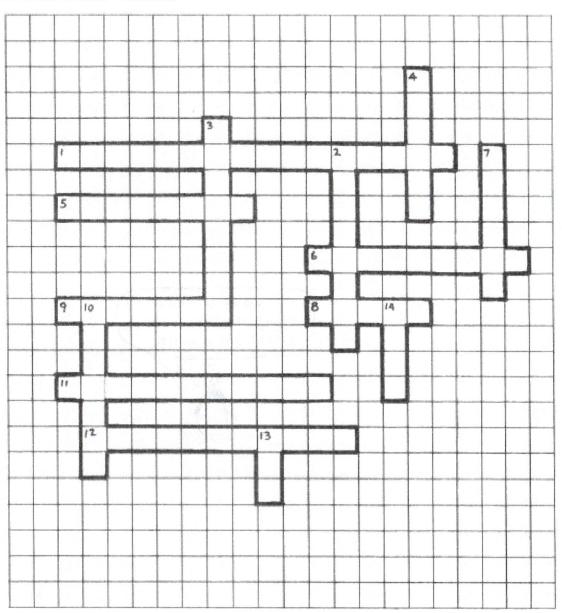

Do geologists who wonder have rocks in their heads?

The five Es of education:
Engage, Explore, Explain,
Extend, and Evaluate.

Conceptual Integrated Science —— Third Edition

Chapter 24: Earth's Surface—Land and Water
Fault Facts

Imagine you are tiny like Perky the mouse. If you fell into a fault during an earthquake,
what would it look like? How would the slipping blocks of rock move?
Answer: It depends on the fault. There are three basic kinds of faults. Rock moves differently
along each fault type. Explore fault types here and in Section Integrated Science (IS) 24A of your textbook.

Normal Fault
Faults are classified by how the two rocky blocks on either side of the fault move relative to one
another. In a normal fault, rock on one side of the fault drops *down*, relative to the other side.
Investigate the diagram now.

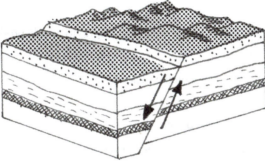

Circle the correct answers:

Do you see that one block of rock looks something like a foot (well, maybe a hoof)? That's the

(hanging wall)   (footwall).

The block of rock that looks like it's hanging over the footwall is called the

(hanging wall)   (footwall).

In a normal fault, the (hanging wall)   (footwall) moves down relative to the

(hanging wall)   (footwall).

Normal faults occur when the stress acting on the rocks is

(tension)   (compression).

For this reason, normal faults are associated with

(convergent)   (divergent)   (transform) plate boundaries.

┌─ **Conceptual Integrated Science** ─ **Third Edition**

Chapter 24: Earth's Surface—Land and Water
Fault Facts—continued

Reverse Fault

Along a reverse fault, one block of rock is pushed *up* relative to the other side. This is the reverse of what gravity would do, correct?

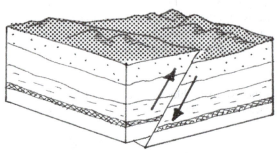

Circle the correct answers:

When movement along a fault is the reverse of what normal gravity would produce, we call the

fault a (normal)   (reverse) fault.

Reverse faults occur when the stress acting on the rocks is

(tension)   (compression).

For this reason, normal faults are associated with

(convergent)   (divergent)   (transform) plate boundaries.

Strike-slip Fault

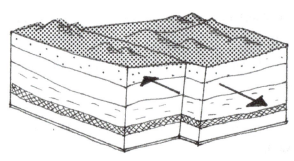

Circle the correct answer:

Blocks of rock move up or down along normal and reverse faults. But strike-slip faults are different. The gigantic blocks of rock on either side of a strike-slip fault scrape beside one another. Motion is horizontal, not vertical. Strike-slip boundaries occur at

(convergent)   (divergent)   (transform) plate boundaries.

How Faults Affect the Strength of Earthquakes

Earthquakes generally occur at faults. The strength of an earthquake depends on the kind of faults involved. Fill in the blanks below showing how fault type relates to earthquake strength.

Plate boundary	Major fault type	Earthquake strength
transform	strike-skip fault	_____
convergent	_____	strong
_____	normal fault	weak

Conceptual Integrated Science
Third Edition

Chapter 24: Earth's Surface—Land and Water
The Story of Old, Folded Mountains—The Appalachians

Folded mountains form when blocks of rock are squeezed together and push upward like wrinkled cloth. Usually folded mountains occur at convergent plate boundaries, where tectonic plates collide. The Appalachian Mountains, however, lie in eastern North America in the middle of the North American plate. How can this be? *Find out how the Appalachian Mountains developed. Then circle the answers on the next page to tell their story.*

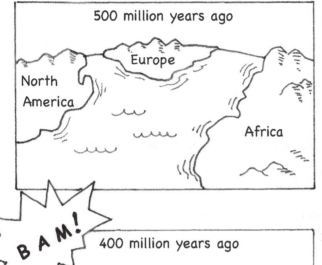

1. The landmasses that would become North America and Africa were moving toward one another about 500 million years ago.

2. About 400 million years ago—BAM! The tectonic plates that North America and Africa were riding on collided. The huge collision caused the crust to fold upward, creating the Appalachian Mountains.

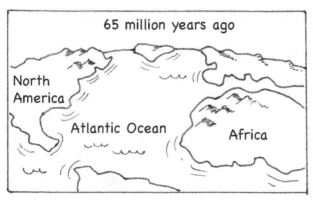

3. About two million years later, North America and Africa began to split. As the Atlantic Ocean grew wider, the Appalachians were shifted away from the plate boundary. By 65 million years ago, the Appalachians had moved to a plate interior. No longer were the Appalachians near a plate boundary.

Conceptual Integrated Science
Third Edition

Chapter 24: Earth's Surface—Land and Water
The Story of Old, Folded Mountains—The Appalachians—continued

Circle the correct answers:

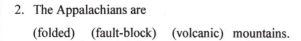

1. The Appalachian Mountains are (old) (young) mountains.

   You can tell this by their low, rounded shapes.

2. The Appalachians are
   (folded) (fault-block) (volcanic) mountains.

3. The forces that produced the Appalachians were due to
   (tension) (compression).

4. About 400 million years ago the Appalachians formed when the landmasses that would be

   North America and Africa (collided) (pulled apart).

5. So, the Appalachian Mountains, like most other folded mountains, developed at a

   (divergent) (convergent) plate boundary.

6. Then, about 200 million years ago, North America and Africa began to break apart.

   The plate boundary that formed between them was (divergent) (convergent).

7. In summary, the Appalachians formed at the boundary of two

   (converging) (diverging) plates. Later, a divergent boundary formed in the Appalachian

   mountain range. This boundary split the mountains. Some of the mountains drifted along with

   Africa and some drifted along with North America. New oceanic lithosphere—the Atlantic

   Ocean basin—was created where the plates were splitting apart. As the seafloor got wider

   over time, the Appalachians were shifted toward the

   (edge) (interior) of the North American Plate. And that's the story of how the Appalachian

   Mountains came to be located in the middle of a

   (tectonic plate) (mid-ocean ridge).

   Are plates still moving?

Experimental evidence is the test of truth in science.

┌─ Conceptual Integrated Science ─ Third Edition

Chapter 24: Earth's Surface—Land and Water

Where the Action Is

1. Investigate the map of the world that shows the main areas where folded mountains exist.

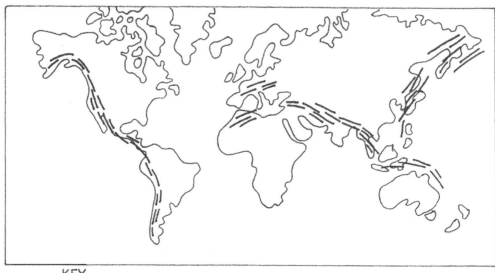

KEY
≡ Main areas of folded mountains

Are folded mountains randomly placed around the globe?

_____

2. Now investigate this map of the world that shows the main areas of volcanic activity and where earthquakes occur.

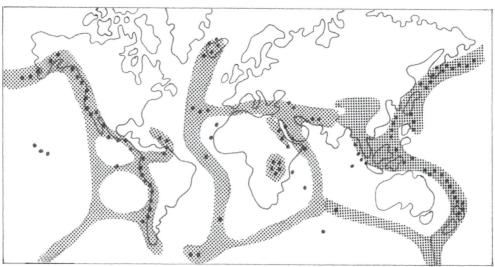

KEY
▨ Region where most earthquakes happen
• Volcano

Conceptual Integrated Science
Third Edition

Chapter 24: Earth's Surface—Land and Water
Where the Action Is—continued

    a. Do earthquakes and volcanic activity generally occur in the same places? _____

    b. If so, why would this be true?

    _____

    _____

3. Compare the two maps on the facing page.

    a. Can you see the pattern? _____

    b. Briefly give an explanation for the pattern.

    _____

    _____

4. Now investigate the map that shows the main tectonic plates. Relate the location of folded mountains, earthquakes, and volcanic activity to tectonic plate boundaries.

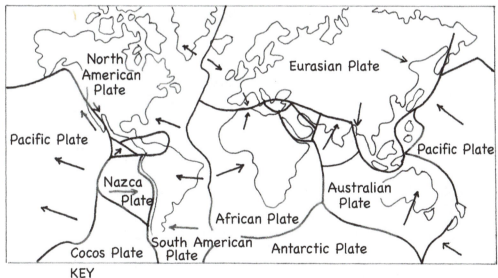

KEY
  ⟋ Plate boundary
  ⟶ Direction that plate is moving

What pattern do you see?

_____

_____

Conceptual Integrated Science Third Edition

Chapter 24: Earth's Surface—Land and Water

Where the Action Is—continued

5. Explain why folded mountains, earthquakes, and volcanic activity are most common near tectonic plate boundaries.

   _____

   _____

   _____

   _____

   _____

   _____

6. Why was there little understanding of how Earth's surface features developed before the 1960s? [Hint: When was the theory of Plate Tectonics developed?]

   _____

   _____

   _____

Most of geology's action is in slow motion.

Conceptual Integrated Science
Third Edition

Chapter 24: Earth's Surface—Land and Water
What Do You Know About Water Flow?

Look at the diagram of the water cycle. Next to each reservoir, write how long water stays in that reservoir.

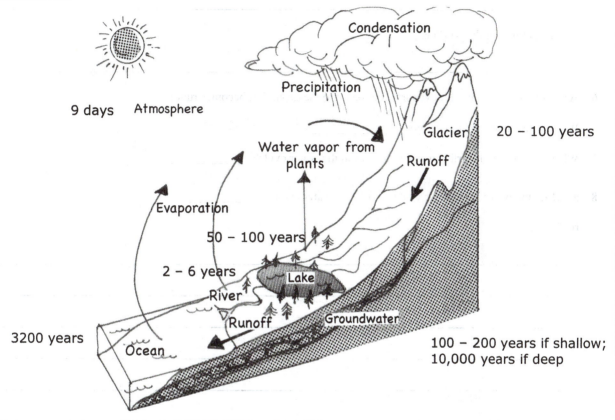

9 days   Atmosphere

Condensation

Precipitation

Glacier   20 – 100 years

Water vapor from plants

Runoff

Evaporation

50 – 100 years

2 – 6 years

Lake

River

Runoff   Groundwater

3200 years   Ocean

100 – 200 years if shallow;
10,000 years if deep

1. What is the source of energy that drives the water cycle? _____

2. a. Where does most of Earth's water reside?

   _____

   b. How long, on average, does it stay there? _____

3. How long does an average water molecule remain in the gaseous state if it completes one full trip through the water cycle beginning and ending in the ocean as shown?

   _____

4. What percentage of Earth's water is fresh? _____

Conceptual Integrated Science
Third Edition

Chapter 24:   Earth's Surface—Land and Water
What Do You Know About Water Flow?—*continued*

5. a. In what form is most of Earth's fresh water?

_____

b. In what part of the world is it located?

_____

6. When precipitation doesn't evaporate or soak into the ground, it becomes runoff.

Where does runoff eventually go?   _____

7. What percentage of Earth's water takes part in the water cycle?   _____

8. a. Does the water in your body take part in the water cycle?   _____

b. Explain.

_____

_____

_____

_____

_____

_____

_____

9. On a typical day in the United States, about 4 trillion gallons of precipitation falls.
    What happens to this water once it falls?

_____

_____

_____

Bottled water costs as much and often more than fruit juice?

Conceptual Integrated Science
Third Edition

Chapter 25: Surface Processes
Weathering Earth's Crust

Weathering is the process by which rocks in Earth's crust are broken down into smaller pieces. These pieces vary in size from boulders to pebbles to the tiny particles that make up soil.

Complete the concept map by filling in the blanks:

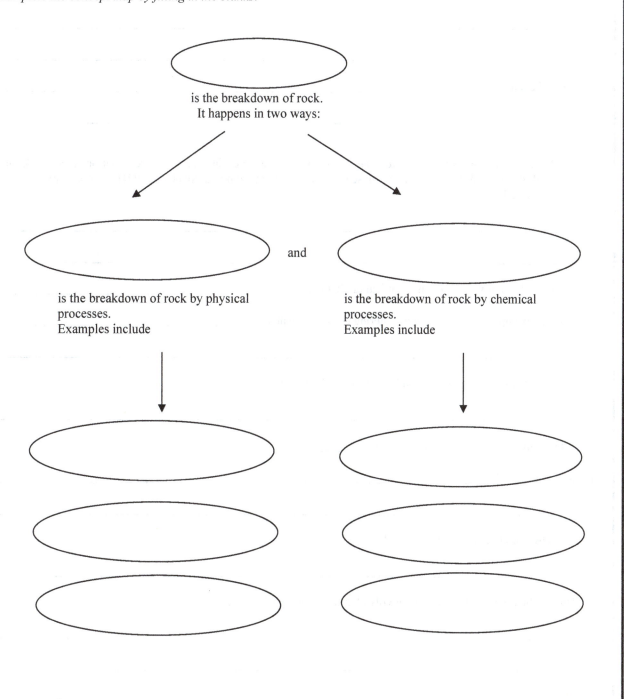

is the breakdown of rock.
It happens in two ways:

and

is the breakdown of rock by physical processes.
Examples include

is the breakdown of rock by chemical processes.
Examples include

Conceptual Integrated Science — Third Edition

Chapter 25: Surface Processes

Weathering Earth's Crust—continued

1. Give two examples of weathering you have observed at home or at school over the past week. Describe how the weathering occurred for each example.

First example: _____

_____

_____

Second example: _____

_____

2. a. Rocks expand and contract due to changes in temperature. Over time, temperature variations crack rocks and break them up. Where else might you see weathering due to temperature changes? [Hint: You walk on it everyday.]

_____

_____

 b. Is this kind of weathering mechanical or chemical? _____

3. Give an example of *mechanical* weathering caused by water. _____

_____

4. Give an example of *chemical* weathering caused by water.

_____

5. What is the final product of weathering? _____

6. a. What is soil? _____

_____

 b. Why does it take so long to form?

_____

7. How is the process of weathering involved in the growing of crops?

_____

_____

_____

Conceptual Integrated Science
Third Edition

Chapter 25: Surface Processes
The Speed of Water Affects Sedimentary Rock Formation

Some rocks form when rock fragments are squeezed together under pressure—we say the rock
is *compacted*. Three common sedimentary rocks formed this way are shown in the table. Note
that shale is made of tiny particles, sandstone is made of particles the size of sand grains,
and conglomerate is made of much bigger particles.

Rock Data

Type of Rock			
	Conglomerate	Sandstone	Shale
Grain Size	larger than 2 mm	between 2 mm and 0.05 mm	less than 0.02 mm

1. Investigate the diagram of water flowing from a river to the sea. Note that water generally slows down as it flows
   toward the sea. Also note that fast-flowing water deposits large grains and slow flowing water deposits small grains.

   _____Flow of water_____ ➤

   _____Water slows down_____ ➤

   _____Size of deposits decreases_____ ➤

In each blank write *conglomerate*, *sandstone*, or *shale* to tell which type of rock forms in each
sedimentary environment.

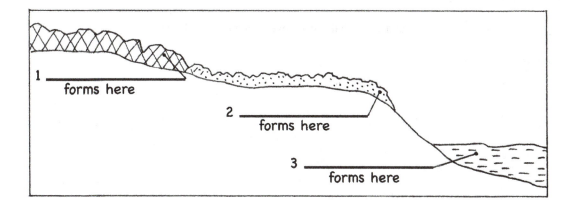

1 _____
   forms here

2 _____
   forms here

3 _____
   forms here

Chapter 25: Surface Processes

The Speed of Water Affects Sedimentary Rock Formation—continued

Fill in the blanks by choosing the correct answer:

2. The faster water flows, the more energy it has and the bigger the particles it can carry.

   Water generally moves

   _____ (faster, slower)   as it flows from steep mountains toward flatter land

   near the sea. As the water slows down, the particles it first deposits are

   _____ (bigger, smaller).

   A sedimentary rock made of large particles, such as stones and pebbles, is

   _____ (conglomerate, sandstone, shale).   So this rock forms along river

   beds where rushing water has lost enough of its speed to begin dropping its larger sediments.

   Water tends to slow down near the mouth of a river, when the river meets the sea. So water

   deposits medium-sized grains, such as sand, near a river's mouth. So near the mouth of a

   river we find

   _____ (conglomerate, sandstone, shale).

   But the smallest particles, such as clay, aren't deposited until they reach the sea. In the sea,

   where water has little energy of motion, small particles are finally deposited. Fine sediments

   are composed of rocks such as

   _____ (conglomerate, sandstone, shale) that develop along the still sea

   bottom over geologic time.

Earth Science is cool

Name _____ Date _____

Chapter 25: Surface Processes
Landforms Created by Erosion

1. a. Label each landform by filling in the blanks.

Delta
Floodplain
Headwaters
Levee
Meander
Mouth
Tributary
V-shaped valley

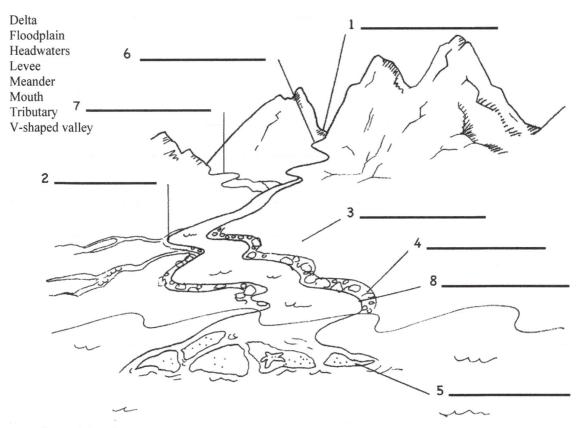

1 _____

6 _____

7 _____

2 _____

3 _____

4 _____

8 _____

5 _____

b. Describe each landform.

Delta _____

_____

_____

_____

Floodplain _____

_____

_____

_____

_____

Name _____ Date _____

Chapter 25: Surface Processes
Landforms Created by Erosion—continued

Headwaters _____

_____

Levee _____

_____

_____

_____

Meander _____

_____

_____

_____

Mouth _____

_____

_____

Tributary _____

_____

V-shaped valley _____

_____

_____

Running water: Earth's most important agent of erosion.

Conceptual Integrated Science — Third Edition

Chapter 26: Weather

Layers of the Atmosphere

Label each of the layers of the atmosphere in the diagram (not drawn to scale).

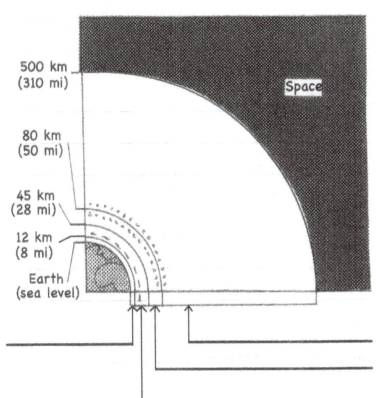

500 km
(310 mi)

80 km
(50 mi)

45 km
(28 mi)

12 km
(8 mi)

Earth
(sea level)

Space

Remember:
Earth's troposphere
is very thin. Its
thickness compared
to Earth is like the
thickness of apple
skin on an apple!

1. In which layer of the atmosphere do we live? _____

2. In which layer does weather occur? _____

3. a. What layer is best for airplane flight? _____

   b. Why? _____

4. What caused the amount of oxygen in Earth's atmosphere to increase when Earth was young? [Hint: What pumps vast amounts of oxygen into the atmosphere daily?]

   _____

   _____

5. Name two characteristics of the atmosphere that make life possible on Earth's surface.

   _____

   _____

Chapter 26: Weather
Layers of the Atmosphere—continued

6. a. What is the ozone layer?

_____

   b. Where is it located? _____

   c. Mark it on the diagram with a dashed line.

   d. What does the ozone layer do for you? _____

_____

_____

7. What is the ionosphere?

_____

_____

   b. Where is it located? _____

   c. Mark it on the diagram with a dotted line.

8. Does temperature increase or decrease with altitude in the troposphere? _____

   b. What is the reason?

_____

_____

9. What is wrong with the picture of the layers of the atmosphere? [Hint: Look at the scale.
    Can the layers of the atmosphere be drawn to scale on a diagram of this size?]

_____

_____

_____

_____

_____

_____

Conceptual Integrated Science — Third Edition

Chapter 26: Weather
Prevailing Winds—How Do They Work?

Fill in the blanks to show where the following prevailing winds or "global winds" are located:
Polar easterlies, Westerlies; Northeast tradewinds; Southeast tradewinds.

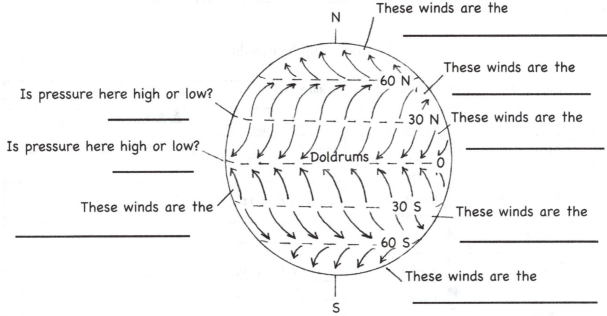

These winds are the

_____

These winds are the

_____

These winds are the

_____

Is pressure here high or low?

_____

Is pressure here high or low?

_____

These winds are the

_____

These winds are the

_____

These winds are the

_____

Circle the correct answers:

1. Global winds are part of a huge pattern of (water) (air) circulation around the globe.

2. Global winds transport heat around the globe, which affects
   (temperature) (solar energy).

3. There are (two) (four) (six) wind belts on Earth. Within each belt, warm air rises, moves
   laterally, then sinks. Wind belts are (convection cells) (zones of high pressure).

4. The turning or spiraling of global winds is due to Earth's (tilt) (rotation).

5. Like all winds, global winds flow from areas of (high) (low) pressure to regions of
   (high) (low) pressure.

6. Earth's atmosphere has regions of high and low pressure because of uneven
   (heating) (rotating) of Earth.

7. Global winds drive ocean (currents) (waves)—circulating streams of water that
   redistribute heat throughout the globe.

8. Warm air rises at the (poles) (equator) and sinks at the (poles) (equator).

Conceptual Integrated Science **Third Edition**

Chapter 26: Weather
Read a Weather Map

Weather Symbols

To report the weather in shorthand, weather symbols are used. Read the weather symbols on the chart below.

Weather Map Symbols

Weather Conditions	Cloud Cover	Wind Speed (mph)	Special Symbols
Light Rain	No Clouds	Calm	Cold Front
Moderate Rain	One-Tenth or Less	3–8	Warm Front
Heavy Rain	Two- to Three-Tenths	9–14	H High Pressure
Drizzle	Broken	15–20	L Low Pressure
Light Snow	Nine-Tenths	21–25	Hurricane
Moderate Snow	Overcast	32–37	
Thunderstorm	Sky Covered	44–48	
Freezing Rain		55–60	
Haze		66–71	
Fog			

Use the above chart to answer the following questions:

1. a. What is the symbol for light snow? _____

   b. What is the symbol for a sky covered by clouds? _____

   c. What is the symbol for winds between 55 and 60 miles per hour? _____

Name _____ Date _____

Conceptual Integrated Science — Third Edition

Chapter 26: Weather

Read a Weather Map—continued

Station Models

Weather data collected at weather stations can be represented on a weather map. Each weather station uses a station model—a summary of the weather that is based on weather symbols. Notice that the station model for Portland shows the city's weather at a certain time.

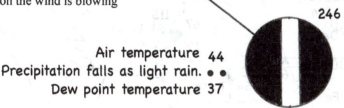

Station Model for Portland, Oregon

Wind speed is shown by whole and half tails. The line shows the direction the wind is blowing from.

Air temperature 44
Precipitation falls as light rain. ● ●
Dew point temperature 37

246

Shading represents cloud coverage. This symbol shows there is a 90% cloud coverage of the sky.

Atmospheric pressure in units of millibars (mbar). The number on the station model is a code. To convert the code to the true pressure, you must follow two rules:

1. If the first code number is over 5,
   place a 9 in front of the code number and
   a decimal point between the last two digits.

2. If the first code number is less than or equal to 5,
   place a 10 in front of the number and
   a decimal point between the last two digits.

Use the above station model for Portland, Oregon, to answer the following questions:

2. a. What was the air temperature at the weather station in Portland on the day the data was recorded? _____

   b. What was the wind speed? _____

   c. How much cloud coverage was there? _____

   d. What was the atmospheric pressure? _____

─ **Conceptual Integrated Science** ─ **Third Edition**

Chapter 26: Weather
Read a Weather Map—continued

Weather Maps

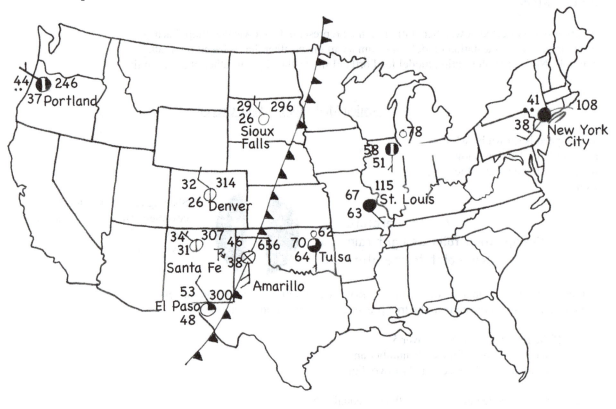

1. Describe the weather in each of these cities. Include all the information on each station model.

   a. Sioux Falls

   _____

   _____

   b. Amarillo

   _____

   _____

   c. New York City

   _____

   _____

—Conceptual Integrated Science Third Edition

Chapter 26:   Weather
Read a Weather Map—continued

2. The arrows on the cold front symbol show the cold front is moving and in what direction. What is the direction in which the cold front is moving?

   _____

3. How is the cold front affecting the weather? (Compare the regions ahead of the front with the locations the front has recently passed through.)

   _____

   _____

   _____

   _____

Whenever you lose your way, run slower—not faster!

For beginners, wisdom is knowing not to throw a rock straight up. Advanced wisdom is knowing what to overlook.

Conceptual Integrated Science
Third Edition

Chapter 27: Environmental Geology
Earthquakes: How Big and How Often?

Earthquake Magnitude and Frequency

Richter Scale Magnitude	Earthquake Effects	Average Number Per Year
less than 2.0	Cannot be felt	600,000+
2.0 to 2.9	Recorded but cannot be felt	300,000
3.0 to 3.9	Felt by most people near epicenter	49,000
4.0 to 4.9	Minor shock; slight damage near epicenter	6,000
5.0 to 5.9	Moderate shock; energy released equals the energy released by one atomic bomb	1,000
6.0 to 6.9	Large shock; damaging to population centers	120
7.0 to 7.9	Major earthquake with severe property damage; can be detected around the world	14
8.0 to 8.9	Great earthquake; communities near epicenter are destroyed. Energy released is equivalent to that of millions of atomic bombs.	once every 5 to 10 years
9.0 to 9.9	Large earthquake recorded	1 to 2 per century

1. What does the Richter scale measure?

   _____

2. About how many earthquakes occur each year but are not felt by people?

   _____

3. How many earthquakes of magnitude 5.0 to 5.9 does it take to release the same amount of energy released by one 8.0 to 8.9 earthquake?

   _____

   _____

Conceptual Integrated Science
Third Edition

Chapter 27: Environmental Geology
Earthquakes: How Big and How Often?—continued

4. a. According to the chart on the previous page, what is a "major earthquake"?

   _____

   _____

   b. What is the range of Richter magnitudes of a major earthquake?

   _____

   c. How many earthquakes occur each year?

   _____

5. Where do most of the world's earthquakes occur?

   _____

   _____

6. Suppose you are near an earthquake epicenter. You can feel the quake but there is no damage. What is the magnitude of the earthquake?

   _____

   _____

7. a. Would you rather endure ten magnitude 3.0 earthquakes or one 8.0 earthquake?

   _____

   b. Why?

   _____

   _____

   _____

Earth quakes, shakes, rocks, and rolls—what a ride!

The process called science replaces *confusion* with *understanding* in a manner that's precise, predictive, and reliable — while providing an empowering and emotional experience.

Conceptual Integrated Science
Third Edition

Chapter 27: Environmental Geology
Is Your Knowledge Shaky?

Use the clues below to fill in the words of the crossword puzzle in the numbered blanks on the facing page.

Across

1. Most earthquakes occur at _____.

5. Force that builds up in rock prior to an earthquake

6. What is released in the form of waves in an earthquake?

9. _____ waves are triggered by rocks slipping at an earthquake focus.

11. The Ring of _____ is a region surrounded by convergent plate boundaries where frequent earthquakes occur.

Down

2. Point at Earth's surface directly above an earthquake focus.

3. The place inside Earth where rock slips, starting an earthquake.

4. A crack in the crust where blocks of rock have moved relative to one another.

7. The earthquake scale that measures shaking of the ground.

8. An instrument that measures earthquake magnitude on the Richter scale.

10. What force holds blocks of strained rock together before an earthquake?

I like puzzles.

Chapter 27: Environmental Geology

Is Your Knowledge Shaky?—continued

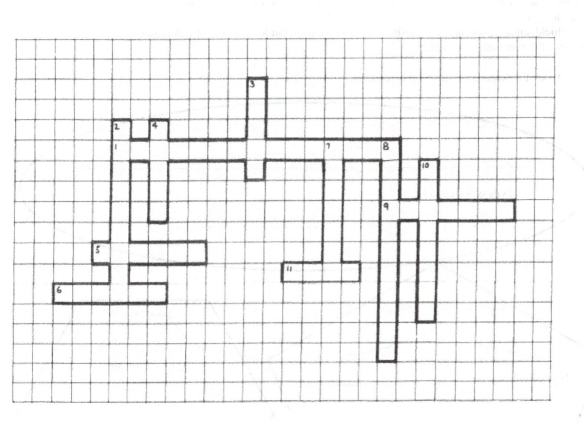

Who was the Richter Scale named after?

Maybe Charles Scale?

┌─Conceptual Integrated Science─ Third Edition

Chapter 27: Environmental Geology
Volcano Varieties

Volcanoes are mountains built up of erupted rocky debris, including lava, ash, and rocky
fragments called *pyroclastics*. Each drawing here represents one of the three types of
volcanoes—shield, cinder cone, and composite. Shield volcanoes, cinder cones, and composite
volcanoes have different sizes and shapes.

1. Label the type of volcano each drawing represents.

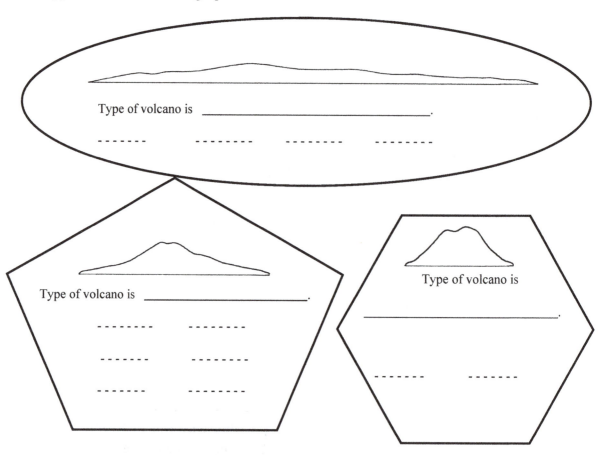

Type of volcano is _____.

- - - - - - -   - - - - - - -   - - - - - - -   - - - - - - -

Type of volcano is _____.

- - - - - - -   - - - - - - -

- - - - - - -   - - - - - - -

- - - - - - -   - - - - - - -

Type of volcano is

_____.

- - - - - - -   - - - - - - -

2. On the dashed lines above, write the letter (A to L) that is appropriate for the type of
volcano.

A. Also called a "stratovolcano"   B. Small but steep volcanoes   C. Tall, broad volcanoes

D. Built from layers of lava,   E. Can trigger *lahars*   F. Kilauea is an example
   ash, and pyroclastics

G. Built up from ash and   H. Gently sloping volcanoes   I. Built from cooled, runny
   cinders   lava

J. Mt. St. Helens is an example   K. Usually erupts quietly   L. Often erupt explosively

Conceptual Integrated Science
Third Edition

Chapter 27: Environmental Geology
Atmospheric Carbon Dioxide and Global Temperature: A Match?

Scientists have been able to estimate Earth's average temperature as well as the concentration of carbon dioxide in Earth's atmosphere over the past several hundred thousand years. The graph below shows this.

Temperature and Carbon Dioxide Concentration Over 160,000 Years

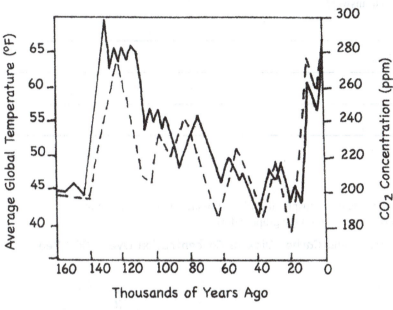

KEY
- Carbon dioxide concentration
- Average global temperature

Can you interpret the graph to answer these questions?

1. Over the past 160,000 years, when was Earth warmest?

_____

2. When Earth reached its highest average global temperature, was atmospheric carbon dioxide also very high?

_____

3. Do the graphs show a *correlation* (connection) between atmospheric carbon dioxide and Earth's average temperature?

_____

_____

Conceptual Integrated Science
Third Edition

Chapter 27: Environmental Geology
Atmospheric Carbon Dioxide and Global Temperature: A Match?—continued

4. a. Were all the variations in atmospheric carbon dioxide caused by human activities?

    _____

    b. What is the reason for your answer?

    _____

    _____

    _____

    _____

Scientists have carefully collected data on atmospheric carbon dioxide and temperature for
the past 1,000 years. This data is represented on the graph below.

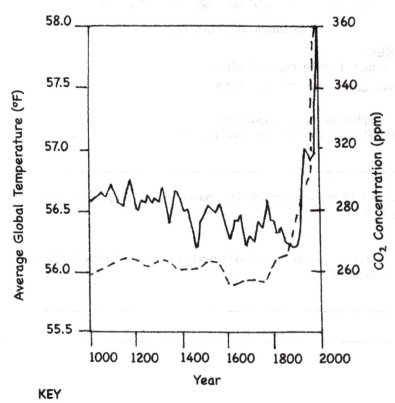

Temperature and Carbon Dioxide Concentration Over 1,000 Years

KEY
/ Carbon dioxide concentration
/ Average global temperature

┌─ **Conceptual Integrated Science** ─ Third Edition

Chapter 27: Environmental Geology
Atmospheric Carbon Dioxide and Global Temperature: A Match?—continued

Use the graph of temperature and carbon dioxide concentration over 1,000 years on the facing page to answer the questions.

5. a. What was Earth's average temperature in 1000 C.E.? _____

    b. What was it in 1900? _____

    c In the year 2000? _____

6. Do the graphs show a *correlation* between atmospheric carbon dioxide and Earth's average temperature since 1860?

_____

_____

7. Why did carbon dioxide levels in the atmosphere begin to rise dramatically around 1860?

_____

_____

_____

8. a. Does the graph show that increased levels of carbon dioxide in the atmosphere *cause* global warming?

_____

    b. Explain.

_____

_____

_____

_____

_____

I'd rather hang out with friends who have reasonable doubts than ones who are absolutely certain about everything.

Conceptual Integrated Science
Third Edition
Update the Law on Climate Change

Congratulations, you have won the election! You are now a local, state, or federal lawmaker. Your first priority is to draft a law on climate change. Your law can specify a policy change to either mitigate climate change or help the public adapt to it. You will need to follow the process below to draft your new law and get it passed.

1. Do Your Research

Review an environmental law to get a sense of how laws are written. Research online or through your local library or town hall to find an environmental law, or *statute*, that you can understand. Note that environmental laws generally include the following:
- Statement of why the law is needed
- Definition of terms
- Statement of who is affected and who is exempt
- Description of how the law will be enforced
- Statement of penalties to any parties violating the law
- Authorization of funding

2. Formulate Your Policy Idea

What kinds of laws are needed in an era of changing climate? Formulate your idea for a new law. Write your statement of intent below. That is, describe the proposed policy change and explain how it will help society mitigate or adapt to climate change. (Include blank lines for a short paragraph.)

3. Define the Scale of Change

Is the change you are proposing a local one that would best be handled by the local authorities? Or does your law involve state-level change, so it needs to be implemented by the state legislature? Or is your law national in scope? If so, it is *federal* legislation that must be handled at the federal level—by Congress.

Decide if you are playing the role of a local, state, or federal lawmaker and write it in the blank below. (Include blank for a few words.)

4. Be Practical

Now that you have a basic idea for your law and its scale, refine it. Review the following practical considerations. Write your answers to these questions below.
a) Does your law have loopholes? How can you close them?
b) How will your law affect jobs and the economy?
c) How much will it cost to implement your law? Where will the money come from?
d) Is the law fair? Is it biased to favor particular groups? Does it impact any demographic group in an overly burdensome manner? (Include a couple of blank lines for parts a–d.)
e) Cite relevant scientific sources to show that the change you propose is feasible in terms of available technology. Also include an environmental impact statement. (Gathering this information for actual legislation would be an exhaustive practice. For the purposes of this assignment, you can limit yourself to 3–5 scientific sources.)

5. Draft Your Law

It's time to roll up your sleeves and draft your law. Write it on a separate piece of paper. It should be detailed and include provisions to make it doable, fair, and effective. (Include blank lines for a short paragraph.)

6. Assess Your Impact

Work with a partner. Let your partner read your law while you read theirs. Write an assessment of the law your partner drafted. Include the following points: 1) What is the most constructive aspect of the law? 2) What could be the short-term impacts of the law? 3) What would be the long-term impacts? 4) Would you vote for or against the law as it is currently written? 5) How could the law be amended so you would vote for it?

7. Maximum Impact: Work On Federal Legislation

Work in small groups of four to six students. Choose group members so that at least one group member has proposed a new federal law. Your group will model the process of taking proposed legislation through committee to be voted on by the full U.S. Senate or House of Representatives. Your group will play the role of a Senate or Congressional committee. First step: Review the legislative process described in the box below.

Conceptual Integrated Science — Third Edition

To Change the Law: Debate and Persuade!

Before a proposed federal law or *bill* is passed into law, it must be debated by politicians representing citizens all across the nation. Therefore, the bill will be debated by representatives of opposing political parties who have different political viewpoints but many common interests, values, and beliefs.

  The bill begins in committee, a bipartisan subgroup of congress persons or senators. If it is approved in committee, the bill moves on to the full House of Representatives or Senate. If the bill wins the vote in both the House of Representatives and the Senate, it becomes law.

Now, allow each member of your group who has authored federal legislation to present their law to the committee. The author reads the bill aloud and explains key features and benefits. Appoint one group member to take notes to keep a record of the debate.

8. Make Amendments

Committee members take turns responding to the proposed bill. Your goal is for the committee to debate every important aspect of it. Meeting notes should be thorough.

  Allow committee members to propose amendments to each bill. Vote on proposed amendments to decide whether they should be included. Now have a final vote in your committee. Is the bill ready to submit to the full chamber?

9. The Final Vote

Your entire class will play the role of the full chamber. Select one member from each committee to present all the bills your committee has approved. Allow questions from the full chamber. Now vote!

10. Reflect

Choose one bill that was passed into law. Explain why that bill was successful in getting through the entire process.

Conceptual Integrated Science
Third Edition

Chapter 28: The Solar System
Earth–Moon–Sun Alignment

Here we see a shadow on a wall cast by an apple. Note how the rays
define the darkest part of the shadow, the *umbra,* and the lighter part of
the shadow, the *penumbra*. The shadows that comprise eclipses of
planetary bodies are similarly formed. Below is a diagram of the Sun,
Earth, and the orbital path of the Moon (dashed circle). One position of
the Moon is shown. Draw the Moon in the appropriate positions on the
dashed circle to represent (a) a quarter moon, (b) a half moon, (c) a solar
eclipse, and (d) a lunar eclipse. Label your positions. For c and d, extend
rays from the top and bottom of the Sun to show umbra and penumbra
regions.

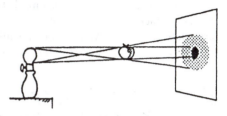

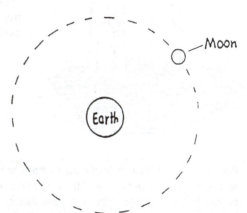

Figure not to scale

The diagram below shows three positions of the Sun: A, B, and C. Sketch the appropriate positions of the Moon in its
orbit about Earth for (a) a solar eclipse and (b) a lunar eclipse. Label your positions. Sketch solar rays similar to the
above exercise.

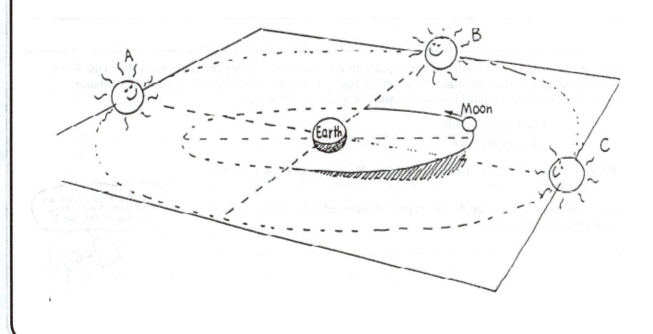

Conceptual Integrated Science
Third Edition

Chapter 28: The Solar System
Pinhole Image Formation

Look carefully at the round spots of light on the shady ground beneath trees. These are *sunballs,* which are images of the Sun. They are cast by openings between leaves in the trees that act as pinholes. (Did you make a pinhole "camera" back in middle school?) Large sunballs, several centimeters in diameter or so, are cast by openings that are relatively high above the ground, while small ones are produced by closer "pinholes." The interesting point is that the ratio of the diameter of the sunball to its distance from the pinhole is the same as the ratio of the Sun's

diameter to its distance from the pinhole. We know the Sun is approximately 150,000,000 km from the pinhole, so careful measurements of the ratio of diameter/distance for a sunball leads you to the diameter of the Sun. That's what this page is about. Instead of measuring sunballs under the shade of trees on a sunny day, make your own easier-to-measure sunball.

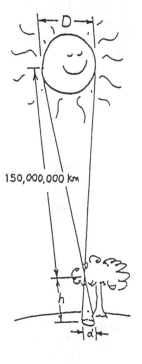

1. Poke a small hole in a piece of card. Perhaps an index card will do, and poke the hole with a sharp pencil or pen. Hold the card in the sunlight and note the circular image that is cast. This is an image of the Sun. Note that its size doesn't depend on the size of the hole in the card, but only on its distance. The image is a circle when cast on a surface perpendicular to the rays—otherwise it's "stretched out" as an ellipse.

2. Try holes of various shapes, say, a square hole or a triangular hole. What is the shape of the image when its distance from the card is large compared with the size of the hole? Does the shape of the pinhole make a difference?

_____

_____

3. Measure the diameter of a small coin. Then place the coin on a viewing area that is perpendicular to the Sun's rays. Position the card so the image of the sunball exactly covers the coin. Carefully measure the distance between the coin and the small hole in the card. Complete the following:

$$\frac{\text{Diameter of sunball}}{\text{Distance to pinhole}} \quad \underline{\hspace{2cm}}$$

With this ratio, estimate the diameter of the Sun. Show your work on a separate piece of paper.

4. If you did this on a day when the Sun is partially eclipsed, what shape of image would you expect to see?

WHAT SHAPE DO SUNBALLS HAVE DURING A PARTIAL ECLIPSE OF THE SUN?

_____

Conceptual Integrated Science
Third Edition

Chapter 28: The Solar System
Jumping Jupiter

Planet X approaches Planet Y in close proximity. What happens next?

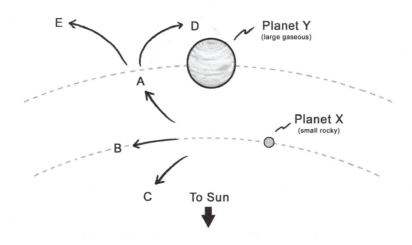

(circle one)

1. Which planet has the greater mass?   Planet X   Planet Y

2. Which planet has the greater orbital speed?   Planet X   Planet Y

3. Five potential paths (A, B, C, D, E) for planet X are indicated with curved arrows. Describe what makes each path possible or not possible.

   A: _____

   B: _____

   C: _____

   D: _____

   E: _____

4. How likely is it for planet X to become a moon with a stable orbit upon following path D?

5. What happens to the orbital velocity of planet X upon following path E?

6. What happens to the distance between the Sun and planet Y upon planet X following path E?

7. What happens to the orbital speed of planet Y upon planet X following path E?

See the Conceptual Academy video tutorial on planet Neptune for further discussions.

Conceptual Integrated Science — Third Edition

Chapter 28: The Solar System

Rings of Saturn

Two small rocks, A and B, are in orbit around a huge gaseous planet.

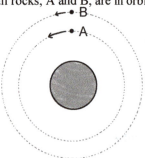 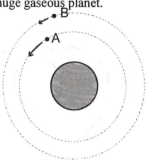 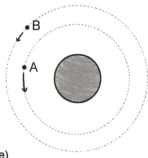

(circle one)

1. Gravity from the planet is weaker for which rock: A   B

2. Which must move sideways faster to remain in orbit: A   B

3. What eventually happens to the distance between them?   _____

A moon made of soft clay is in orbit. Draw arrows to indicate the orbital velocities of the near and far sides of this soft moon.

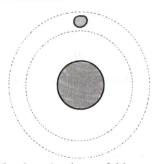 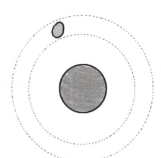 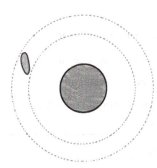

4. Why does the shape of this moon elongate?   _____

5. Might this moon eventually rip apart?   _____

6. If this moon were instead made of iron, electrical forces of attraction between iron atoms would minimize the elongation. But there's another reason elongation would be minimized. What is it?

_____

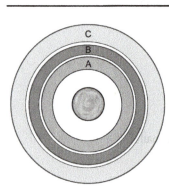

Consider the planetary ring system shown to the left where each ring consists of countless small chunks of water ice.

7. Rank the rings in order of increasing orbital velocity

_____ > _____ > _____

8. Do you suppose the ice chunks within a single ring are each orbiting at the same velocity?

9. Why can't these particles coalesce together into a moon?

10. What would be the fate of Saturn's rings if each particle were made of rock rather than ice?

See the Conceptual Academy video tutorial on planet Saturn for further discussions.

Conceptual Integrated Science — Third Edition

Chapter 28: The Solar System
Word Play

After you have read Chapter 28, use this word play to help solidify some key terms. Only refer back to the textbook after you have given this word play a solid try. Remember, the more you attempt to articulate what you think you understand, the greater the durability of that understanding.

1. A type of nebula having something in common with our atmosphere

2. A prominent constellation located along the celestial equator

3. A slowly rotating cloud of these likely formed our solar system

4. Formed upon the formation of a star

5. The growth of a massive object by gravitational attraction of matter is called

6. A dense interstellar cloud that obscures light

7. Type of red nebula containing hydrogen atoms ionized by nearby stars

8. An intra-galactic cloud of gas and dust

1. __ __ (__) __ __ __ __ __ __ __

2. __ __ (__) __ __

3. __ __ __ __ __ __ __ (__) __ __ __ __ __

4. __ __ __ __ __ __

5. __ __ __ __ __ (__) __ __ __

6. __ (__) __ __

7. __ __ (__) __ __ __ __ __

8. __ __ __ __ __ __

What do you get when you cross a trout with a comet? _____

Thanks to Colin Flanders

┌─**Conceptual Integrated Science**─────────────
    Third Edition

Chapter 28: The Solar System
Solar Anatomy

After you have studied Chapter 28, identify the following structures of our Sun. Try your best to remember these structures before looking back to the chapter for confirmation.

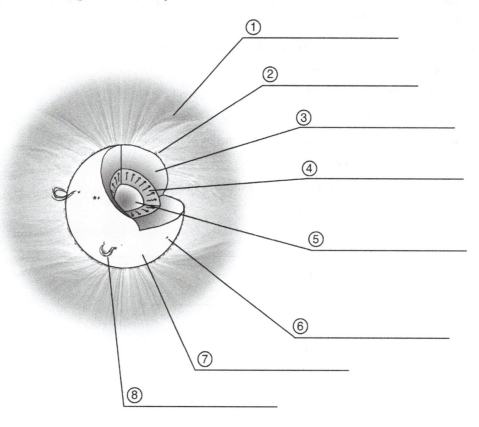

Once you are proficient at recalling the names of the above structures, write down your understanding, in brief, of what happens in structures 3, 4, 5, and 8.

3._____
_____
_____

4._____
_____
_____

5._____
_____
_____

8._____
_____
_____

Conceptual Integrated Science
Third Edition

Chapter 28: The Solar System

Distance to the Moon

It takes the Moon 27.3 days to orbit Earth. We can use this observation to calculate the distance to the Moon. First, convert 27.3 days into seconds using this string of conversion factors:

$$\left(27.3 \text{ days}\right)\left(\frac{\underline{\hspace{1cm}} \text{ hours}}{\underline{\hspace{1cm}} \text{ days}}\right)\left(\frac{\underline{\hspace{1cm}} \text{ min}}{\underline{\hspace{1cm}} \text{ hours}}\right)\left(\frac{\underline{\hspace{1cm}} \text{ sec}}{\underline{\hspace{1cm}} \text{ min}}\right) = \underline{\hspace{3cm}} \text{ sec}$$

In this many seconds, the Moon travels a circumference, which equals 2πr, where "r" is the radius, which is the distance between the centers of Earth and the Moon. This distance traveled over seconds is the Moon's orbital velocity:

$$\frac{\text{distance}}{\text{time}} = \frac{\text{circumference}}{\text{time}} = \left(\qquad \right) = \text{orbital velocity}$$

Orbital velocity is also given by the equation:

radius, r

Earth   Moon

Where G is the gravitational universal constant, $6.6677 \times 10^{-11} \dfrac{\text{N} \cdot \text{m}^2}{\text{kg}^2}$, and M is the mass of Earth, 5.97×10^{27} kg.

Combine the above two equations to calculate the distance to the Moon, r. Your answer should be around 383,000 km.

Use this space to solve for r

Math Counts!

Conceptual Integrated Science — Third Edition

Chapter 28: The Solar System
Diameter of the Moon

Hold a Ping Pong ball far out enough so that it appears to be the same size of the Moon.
Note how the phase of the ball and the Moon are the same.

 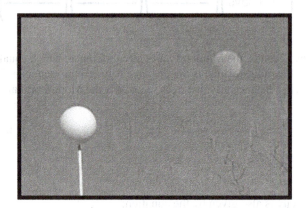

The distance between you and the Ping Pong ball will be about 4.4 meters. The Ping Pong ball itself is 40 centimeters wide, which is 0.040 meters.

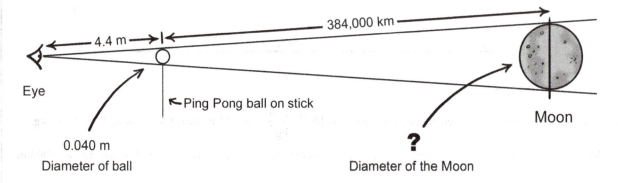

From the previous page we know that the distance to the Moon is about 384,000 km. Use all this information to calculate the diameter of the Moon. Use the following ratio:

Use this area to show your work:

Conceptual Integrated Science
Third Edition

Chapter 29: The Universe
Stellar Parallax

Finding distances to objects beyond the solar system is based on the simple phenomenon of *parallax.* Hold a pencil at arm's length and view it against a distant background—each eye sees a different view (try it and see). The displaced view indicates distance. Likewise, when Earth travels around the Sun each year, the position of relatively nearby stars shifts slightly relative to the background stars. By carefully measuring this shift, astronomer types can determine the distance to nearby stars.

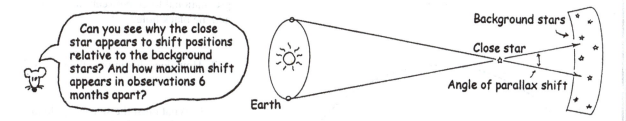

The photographs below show the same section of the evening sky taken at a 6-month interval. Investigate the photos carefully and determine which star appears in a different position relative to the others. Circle the star that shows a parallax shift.

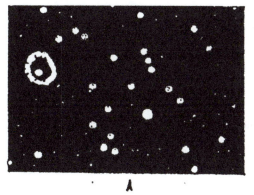

A

B

Below are three sets of photographs, all taken at 6-month intervals. Circle the stars that show a parallax shift in each of the photos.

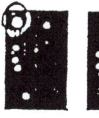

Set A   Set B   Set C

Use a fine ruler and measure the distance of shift in millimeters and place the values below:

Set A _____ mm   Set B _____ mm   Set C _____ mm

Which set of photos indicates the closest star? The most distant "parallaxed" star?

Conceptual Integrated Science — Third Edition

Chapter 29: The Universe
Black Holes

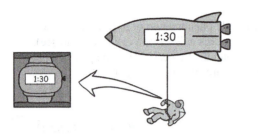

Imagine a ship in orbit around a black hole. An onboard clock reads 1:30 as does the wrist watch of an astronaut about to be lowered toward the black hole.

1. As the astronaut is lowered toward the black hole, what happens to the gravitational force between the astronaut and the black hole?

2. How does this impact the spaceship?

3. What must the spaceship do to maintain orbit?

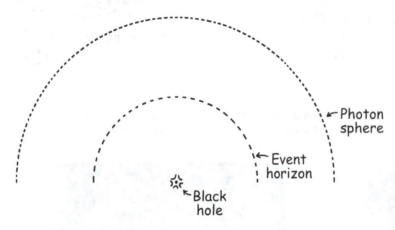

Astronaut's Point of View

4. As the astronaut passes through the photon sphere, what does she notice about the time on her watch?

5. How about when she passes through the event horizon?

6. As the astronaut passes through the photon sphere, what does she notice about the time on the ship's clock?

7. How about when she passes through the event horizon?

Ship's Point of View

8. Assume the astronaut experienced 10 minutes as she was lowered to the photon sphere. How many minutes did her shipmates find it took her to be lowered that far? (circle one)

    a) less than 10 minutes   b) 10 minutes   c) more than 10 minutes

Ship's Point of View (continued)

9. Imagine the tether unexpectedly cuts loose just after the astronaut has passed into the photon sphere. From the ship's point of view, how long does it take for the falling astronaut to reach the event horizon?

10. By the time the astronaut has passed through the event horizon, what has happened to our universe as we know it?

11. Is it possible for an object to pass through a black hole's event horizon?

12. How long does this take?   (circle one)

    (a) From the object's point of view: not long   forever

    (b) From our point of view: not long   forever

Conceptual Integrated Science ── Third Edition

Chapter 29: The Universe
Hertzsprung–Russell Diagram

The Hertzsprung–Russell diagram plots stellar temperature (color) by luminosity. Each dot on this diagram represents a star.

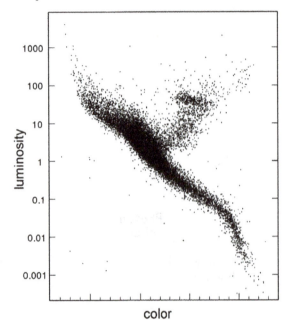

1. Draw an arrow pointing to the largest star on this diagram.

2. Draw a rectangle around the most quickly evolving stars.

3. Draw a triangle over the longest living star.

4. Circle the shortest lived stars.

5. Place an X over a white dwarf.

6. Trace a star shape over the approximate coordinates of our Sun.

7. Draw a large letter B over the bluish stars.

8. Draw a large letter R over the reddish stars.

9. How is the image from an H–Z diagram different from the image of an actual galaxy?

Shown below are H–Z diagrams from five different star clusters of various ages found within our Milky Way Galaxy.

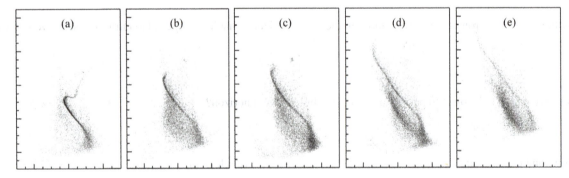

10. Are these star clusters (left to right) presented from old to younger or from young to older? Please explain.

Conceptual Integrated Science
Third Edition

Chapter 29: The Universe
Solar Life Cycle

1. Sort the following frames depicting a star's life cycle in the correct sequence:

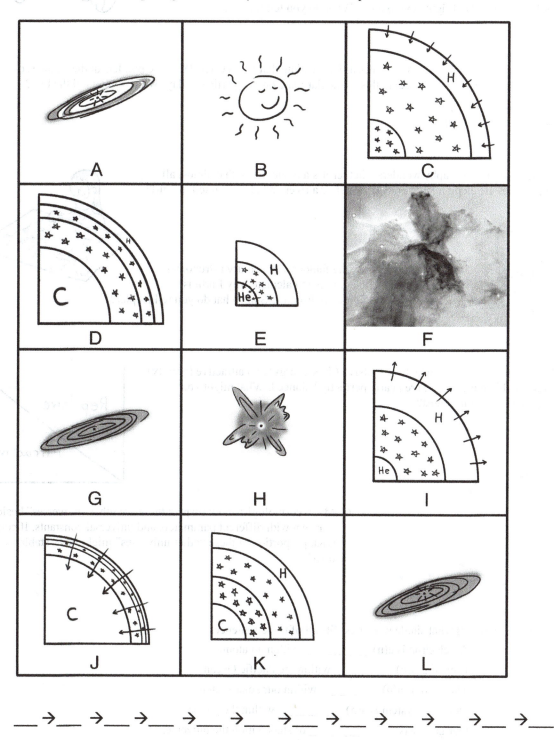

__ → __ → __ → __ → __ → __ → __ → __ → __ → __ → __

2. Circle the blue giant. Draw an arrow pointing to the white dwarf. Draw a triangle over the frame showing a structure of the greatest size.

⌐Conceptual Integrated Science⌐ Third Edition

Chapter 29: The Universe
The Big Picture

1. A chewed apple core lands next to an, who thinks, "How did this beautiful apple happen to land right next to me?" What do you tell the ant?

2. A boy sees a license plate that reads CTX-4872. He ponders at the chances of seeing this particular plate on this particular day. What do you tell the boy?

3. A girl studying geography wonders whether it's a coincidence that almost all major cities are along a large body of water or a river. What do you tell the girl?

4. Johannes Kepler, the famous 17th century astronomer and mathemetician, tries to calculate why Earth is 150,000,000 kilometers from the Sun. What do you tell Kepler?

5. A cosmologist wonders why the repulsive (dark energy) and attractive (gravity) forces within our universe are nearly perfectly balanced. What might you suggest to the cosmologist?

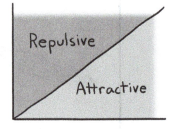

6. Many cosmologists now suspect there are other "universes" besides our own with different parameters and universal constants. If correct, what proportion of these "other universes" might be as stable as our own?

7. Discuss with others what single word best fits all of these sentences:

An electron is a(n) _____ within an atom.

Hawaii is a(n) _____ within the Pacific Ocean.

The Earth is a(n) _____ within our solar system.

Our solar system is a(n) _____ within the galaxy.

Our galaxy is a(n) _____ of stars within the universe.

Our universe is a(n) _____ of galaxies within a larger multiverse.

Conceptual Integrated Science — Third Edition

Chapter 1: About Science
Measuring the Size of Planet Earth

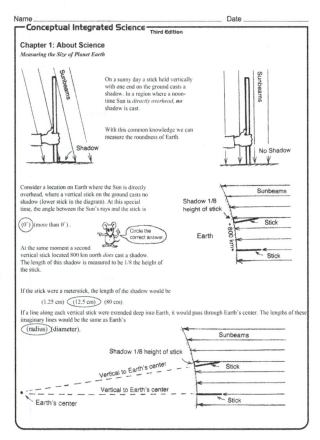

On a sunny day a stick held vertically with one end on the ground casts a shadow. In a region where a noon-time Sun is *directly overhead*, **no** shadow is cast.

With this common knowledge we can measure the roundness of Earth.

Consider a location on Earth where the Sun is directly overhead, where a vertical stick on the ground casts no shadow (lower stick in the diagram). At this special time, the angle between the Sun's rays and the stick is

(0°) (more than 0°).

Circle the correct answer.

At the same moment a second vertical stick located 800 km north *does* cast a shadow. The length of this shadow is measured to be 1/8 the height of the stick.

If the stick were a meterstick, the length of the shadow would be

(1.25 cm) (12.5 cm) (80 cm).

If a line along each vertical stick were extended deep into Earth, it would pass through Earth's center. The lengths of these imaginary lines would be the same as Earth's

(radius) (diameter).

Conceptual Integrated Science — Third Edition

Think ratio and proportion: There are two similar triangles to consider. First, a small one, two sides of which are the northern stick and its shadow. The stick is 8 times as long as its shadow.

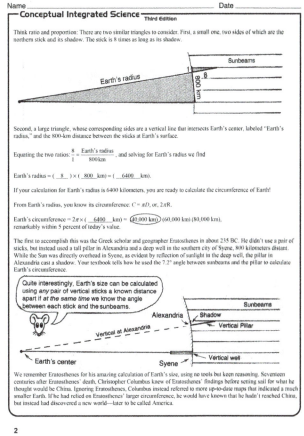

Second, a large triangle, whose corresponding sides are a vertical line that intersects Earth's center, labeled "Earth's radius," and the 800-km distance between the sticks at Earth's surface.

Equating the two ratios: $\frac{8}{1} = \frac{\text{Earth's radius}}{800\,\text{km}}$, and solving for Earth's radius we find

Earth's radius = (8) × (800 km) = (6400 km).

If your calculation for Earth's radius is 6400 kilometers, you are ready to calculate the circumference of Earth!

From Earth's radius, you know its circumference: $C = \pi D$, or, $2\pi R$.

Earth's circumference = $2\pi \times$ (6400 km) = (40,000 km) (60,000 km) (80,000 km), remarkably within 5 percent of today's value.

The first to accomplish this was the Greek scholar and geographer Eratosthenes in about 235 BC. He didn't use a pair of sticks, but instead used a tall pillar in Alexandria and a deep well in the southern city of Syene, 800 kilometers distant. While the Sun was directly overhead in Syene, as evident by reflection of sunlight in the deep well, the pillar in Alexandria cast a shadow. Your textbook tells how he used the 7.2° angle between sunbeams and the pillar to calculate Earth's circumference.

Quite interestingly, Earth's size can be calculated using *any* pair of vertical sticks a known distance apart if *at the same time* we know the angle between each stick and the sunbeams.

We remember Eratosthenes for his amazing calculation of Earth's size, using no tools but keen reasoning. Seventeen centuries after Eratosthenes' death, Christopher Columbus knew of Eratosthenes' findings before setting sail for what he thought would be China. Ignoring Eratosthenes, Columbus instead referred to more up-to-date maps that indicated a much smaller Earth. If he had relied on Eratosthenes' larger circumference, he would have known that he hadn't reached China, but instead had discovered a new world—later to be called America.

Conceptual Integrated Science — Third Edition

Chapter 2: Describing Motion
Vectors and Equilibrium

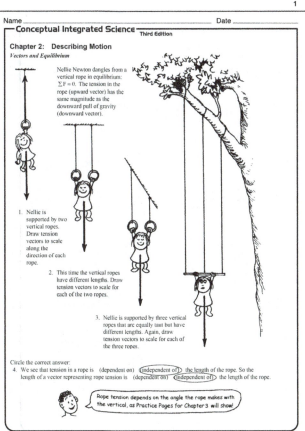

Nellie Newton dangles from a vertical rope in equilibrium: $\Sigma F = 0$. The tension in the rope (upward vector) has the same magnitude as the downward pull of gravity (downward vector).

1. Nellie is supported by two vertical ropes. Draw tension vectors to scale along the direction of each rope.

2. This time the vertical ropes have different lengths. Draw tension vectors to scale for each of the two ropes.

3. Nellie is supported by three vertical ropes that are equally taut but have different lengths. Again, draw tension vectors to scale for each of the three ropes.

Circle the correct answer:

4. We see that tension in a rope is (dependent on) (independent of) the length of the rope. So the length of a vector representing rope tension is (dependent on) (independent of) the length of the rope.

Rope tension depends on the angle the rope makes with the vertical, as Practice Pages for Chapter 3 will show!

Conceptual Integrated Science — Third Edition

Chapter 2: Describing Motion
Free Fall Speed

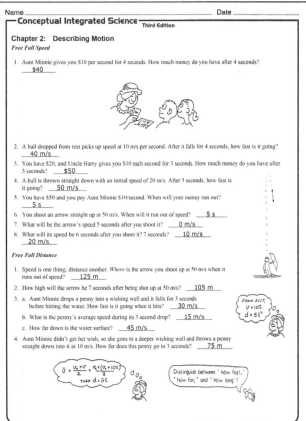

1. Aunt Minnie gives you $10 per second for 4 seconds. How much money do you have after 4 seconds? $40

2. A ball dropped from rest picks up speed at 10 m/s per second. After it falls for 4 seconds, how fast is it going? 40 m/s

3. You have $20, and Uncle Harry gives you $10 each second for 3 seconds. How much money do you have after 3 seconds? $50

4. A ball is thrown straight down with an initial speed of 20 m/s. After 3 seconds, how fast is it going? 50 m/s

5. You have $50 and you pay Aunt Minnie $10/second. When will your money run out? 5 s

6. You shoot an arrow straight up at 50 m/s. When will it run out of speed? 5 s

7. What will be the arrow's speed 5 seconds after you shoot it? 0 m/s

8. What will its speed be 6 seconds after you shoot it? 7 seconds? 10 m/s 20 m/s

Free Fall Distance

1. Speed is one thing; distance another. *Where* is the arrow you shoot up at 50 m/s when it runs out of speed? 125 m

2. How high will the arrow be 7 seconds after being shot up at 50 m/s? 105 m

3. a. Aunt Minnie drops a penny into a wishing well and it falls for 3 seconds before hitting the water. How fast is it going when it hits? 30 m/s

b. What is the penny's average speed during its 3 second drop? 15 m/s

c. How far down is the water surface? 45 m/s

4. Aunt Minnie didn't get her wish, so she goes to a deeper wishing well and throws a penny straight down into it at 10 m/s. How far does this penny go in 3 seconds? 75 m

$$\bar{v} = \frac{v_i + v_f}{2} = \frac{v_i + (v_i + 10t)}{2}$$
$$\text{THEN } d = \bar{v}t$$

Distinguish between "how fast," "how far," and "how long"!

FROM REST,
$v = 10t$
$d = 5t^2$

Acceleration of Free Fall

A rock dropped from the top of a cliff picks up speed as it falls. Pretend that a speedometer and odometer are attached to the rock to show readings of speed and distance at 1-second intervals. Both speed and distance are zero at time = zero (see sketch). Note that after the rock falls 1 second the speed reading is 10 m/s and the distance fallen is 5 m. The readings for succeeding seconds of fall are not shown and are left for you to complete. Draw the position of the speedometer pointer and write in the correct odometer reading for each time. Use $g = 10$ m/s^2 and neglect air resistance.

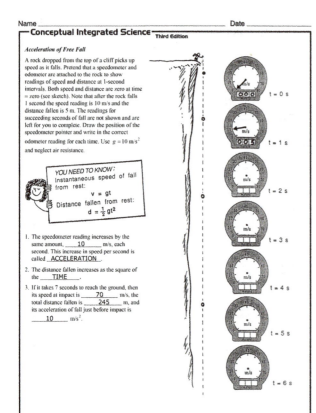

YOU NEED TO KNOW:
Instantaneous speed of fall from rest:
$$v = gt$$
Distance fallen from rest:
$$d = \tfrac{1}{2} gt^2$$

1. The speedometer reading increases by the same amount, __10__ m/s, each second. This increase in speed per second is called __ACCELERATION__.

2. The distance fallen increases as the square of the __TIME__.

3. If it takes 7 seconds to reach the ground, then its speed at impact is __70__ m/s, the total distance fallen is __245__ m, and its acceleration of fall just before impact is __10__ m/s^2.

t = 0 s
t = 1 s
t = 2 s
t = 3 s
t = 4 s
t = 5 s
t = 6 s

6

Chapter 3: Newton's Laws of Motion
Newton's First Law and Friction

1. A crate filled with video games rests on a horizontal floor. Only gravity and the support force of the floor act on it, as shown by the vectors for weight **W** and normal force **N**.
   a. The net force on the crate is (zero) (greater than zero).
   b. Evidence for this is _____ NO ACCELERATION _____

2. A slight pull **P** is exerted on the crate, not enough to move it. A force of friction **f** now acts,
   a. which is (less than) (equal to) (greater than) **P**.
   b. Net force on the crate is (zero) (greater than zero).

3. Pull **P** is increased until the crate begins to move. It is pulled so that it moves with constant velocity across the floor.
   a. Friction **f** is (less than) (equal to) (greater than) **P**.
   b. Constant velocity means acceleration is (zero) (greater than zero).
   c. Net force on the crate is (less than) (equal to) (greater than) zero.

4. Pull **P** is further increased and is now greater than friction **f**.
   a. Net force on the crate is (less than) (equal to) (greater than) zero.
   b. The net force acts toward the right, so acceleration acts toward the (left) (right).

5. If the pulling force **P** is 150 N and the crate doesn't move, what is the magnitude of **f**? __150 N__
6. If the pulling force **P** is 200 N and the crate doesn't move, what is the magnitude of **f**? __200 N__
7. If the force of sliding friction is 250 N, what force is necessary to keep the crate sliding at constant velocity? __250 N__
8. If the mass of the crate is 50 kg and sliding friction is 250 N, what is the acceleration of the crate when the pulling force is 250 N? __0 m/s^2__ 300 N? __1 m/s^2__ 500 N? __5 m/s^2__

7

Nonaccelerated Motion

1. The sketch shows a ball rolling at constant velocity along a level floor. The ball rolls from the first position shown to the second in 1 second. The two positions are 1 meter apart. Sketch the ball at successive 1-second intervals all the way to the wall (neglect resistance.)

   a. Did you draw successive ball positions evenly spaced, farther apart, or close together? Why?
   EVENLY SPACED – EQUAL DISTANCE IN EQUAL TIME → CONSTANT v

   b. The ball reaches the wall with a speed of __1__ m/s and takes a time of __5__ seconds.

2. Table 1 shows the data of sprinting speeds of some animals. Make whatever computations are necessary to complete the table.

Table 1

ANIMAL	DISTANCE	TIME	SPEED
CHEETAH	75 m	3 s	25 m/s
GREYHOUND	160 m	10 s	16 m/s
GAZELLE	1 km	0.01 h	100 km/h
TURTLE	30 cm	30 s	1 cm/s

Accelerated Motion

3. An object starting from rest gains a speed when it undergoes uniform acceleration. The distance it covers is $d = 1/2 \, at^2$. Uniform acceleration occurs for a ball rolling down an inclined plane. The plane below is tilted so a ball picks up a speed of 2 m/s each second; then its acceleration is $a = 2$ m/s^2. The positions of the ball are shown at 1-second intervals. Complete the six blank spaces for distance covered, and the four blank spaces for speeds.

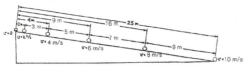

   a. Do you see that the total distance from the starting point increases as the square of the time? This was $v = at$ as discovered by Galileo. If the incline were to continue, predict the ball's distance from the starting point for the next 3 seconds.
   YES; DISTANCE INCREASES AS THE SQUARE OF TIME; 36 m, 49 m, 64 m.

   b. Note the increase of distance between ball positions with time. Do you see an odd-integer pattern (also discovered by Galileo) for the increase? If the incline were to continue, predict the successive distances between ball positions for the next 3 seconds.
   YES; 11 m, 13 m, 15 m

8

Chapter 3: Newton's Laws of Motion
A Day at the Races with Newton's Second Law: $a = \dfrac{F}{m}$

In each situation below, Cart A has a mass of **1 kg**. The mass of Cart B varies as indicated. Circle the correct answer (A, B, or Same for both).

1. Cart A is pulled with a force of **1 N**. Cart B also has a mass of **1 kg** and is pulled with a force of **2 N**. Which undergoes the greater acceleration?

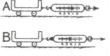

2. Cart A is pulled with a force of **1 N**. Cart B has a mass of **2 kg** and is also pulled with a force of **1 N**. Which undergoes the greater acceleration?

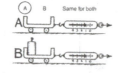

3. Cart A is pulled with a force of **1 N**. Cart B has a mass of **2 kg** and is pulled with a force of **2 N**. Which undergoes the greater acceleration?

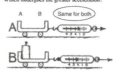

4. Cart A is pulled with a force of **1 N**. Cart B has a mass of **3 kg** and is pulled with a force of **3 N**. Which undergoes the greater acceleration?

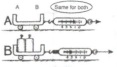

5. This time Cart A is pulled with a force of **4 N**. Cart B has a mass of **4 kg** and is pulled with a force of **4 N**. Which undergoes the greater acceleration?

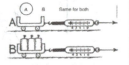

6. Cart A is pulled with a force of **2 N**. Cart B has a mass of **4 kg** and is pulled with a force of **3 N**. Which undergoes the greater acceleration?

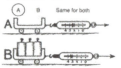

9

282

Conceptual Integrated Science — Third Edition

Chapter 3: Newton's Laws of Motion
Dropping Masses and Accelerating Cart

1. Consider the simple case of a 1-kg cart being pulled by a 10 N applied force. According to Newton's Second Law, acceleration of the cart is

$$a = \frac{F}{m} = \frac{10\ N}{1\ kg} = 10\ m/s^2$$

This is the same as the acceleration of free fall, g—because a force equal to the cart's weight accelerates it.

2. Now consider the acceleration of the cart when a second mass is also accelerated. This time the applied force is due to a 10-N iron weight attached to a string draped over a pulley. Will the cart accelerate as before, at 10 m/s² ? The answer is *no*, because the mass being accelerated is the mass of the cart *plus* the mass of the piece of iron that pulls it. Both masses accelerate. The mass of the 10-N iron weight is 1 kg, so the total mass being accelerated (cart + iron) is 2 kg. Then,

$$a = \frac{F}{m} = \frac{10\ N}{2\ kg} = 5\ m/s^2$$

The pulley changes only the direction of the force.

Don't forget: the total mass of a system includes the mass of the hanging iron.

Note this is half the acceleration due to gravity alone, g. So the acceleration of 2 kg produced by the weight of 1 kg is g/2.

a. Find the acceleration of the 1-kg cart when two identical 10-N weights are attached to the string.

$$a = \frac{F}{m} = \frac{unbalanced\ force}{total\ mass} = \frac{20\ N}{3\ kg} = \underline{6.7}\ m/s^2.$$

Note that the mass being accelerated is 1 kg for the cart + 1 kg each for the weights = 3 kg.

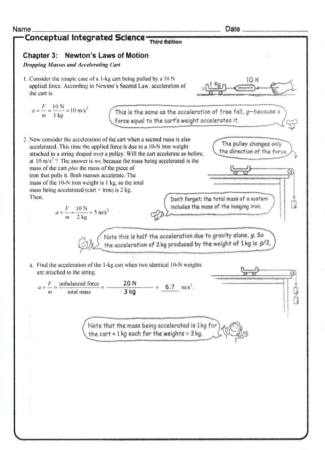

Conceptual Integrated Science — Third Edition

Dropping Masses and Accelerating Cart—continued

b. Find the acceleration of the 1-kg cart when three identical 10-N weights are attached to the string.

$$a = \frac{F}{m} = \frac{unbalanced\ force}{total\ mass} = \frac{30\ N}{4\ kg} = \underline{7.5}\ m/s^2.$$

c. Find the acceleration of the 1-kg cart when four identical 10-N weights (not shown) are attached to the string.

$$a = \frac{F}{m} = \frac{unbalanced\ force}{total\ mass} = \frac{40\ N}{5\ kg} = \underline{8.0}\ m/s^2.$$

d. This time, 1 kg of iron is added to the cart, and only one iron piece dangles from the pulley. Find the acceleration of the cart.

$$a = \frac{F}{m} = \frac{unbalanced\ force}{total\ mass} = \frac{10\ N}{3\ kg} = \underline{3.4}\ m/s^2.$$

The force due to gravity on a mass m is mg. So gravitational force on 1 kg is (1 kg)(10 m/s²) = 10 N.

e. Find the acceleration of the cart when it carries two pieces of iron and only one iron piece dangles from the pulley.

$$a = \frac{F}{m} = \frac{unbalanced\ force}{total\ mass} = \frac{10\ N}{4\ kg} = \underline{2.5}\ m/s^2.$$

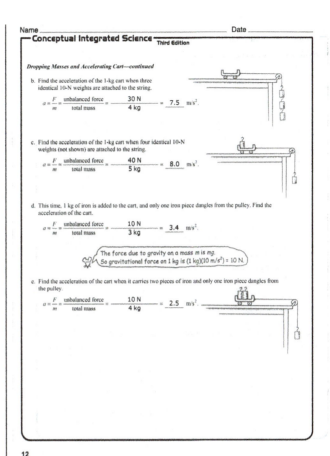

Conceptual Integrated Science — Third Edition

Dropping Masses and Accelerating Cart—continued

f. Find the acceleration of the cart when it carries three pieces of iron and only one iron piece dangles from the pulley.

$$a = \frac{F}{m} = \frac{unbalanced\ force}{total\ mass} = \frac{10\ N}{5\ kg} = \underline{2.0}\ m/s^2.$$

g. Find the acceleration of the cart when it carries three pieces of iron and four iron pieces dangle from the pulley.

$$a = \frac{F}{m} = \frac{unbalanced\ force}{total\ mass} = \frac{40\ N}{8\ kg} = \underline{5.0}\ m/s^2.$$

How does this compare with the acceleration of (f) above, and why?

Mass of cart is 1 kg. Mass of 10-N iron is also 1 kg.

h. Draw your own combination of masses and find the acceleration.

OPEN

$$a = \frac{F}{m} = \frac{unbalanced\ force}{total\ mass} = \underline{\quad\quad} = \underline{\quad}\ m/s^2.$$

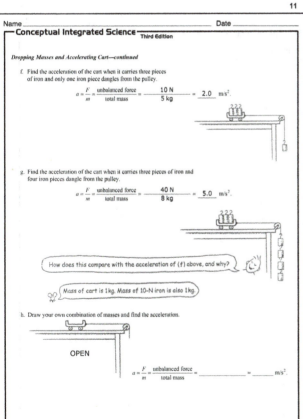

Conceptual Integrated Science — Third Edition

Chapter 3: Newton's Laws of Motion
Mass and Weight

Learning physics is learning the connections among concepts in nature, and also learning to distinguish between closely related concepts. Velocity and acceleration are often confused. Similarly, in this chapter, we find that mass and weight are often confused. They aren't the same! Please review the distinction between mass and weight in your textbook. To reinforce your understanding of this distinction, circle the correct answers below.

Comparing the concepts of mass and weight, one is basic—fundamental—depending only on the internal makeup of an object and the number and kind of atoms that compose it. The concept that is fundamental is (mass) (weight).

The concept that additionally depends on location in a gravitational field is (mass) (weight).

To repeat for emphasis, (Mass) (Weight) is a measure of the amount of matter in an object and only depends on the number and kind of atoms that compose it.

We can correctly say that (mass) (weight) is a measure of an object's "laziness."

(Mass) (Weight) is related to the gravitational force acting on the object.

(Mass) (Weight) depends on an object's location, whereas (mass) (weight) does not.

In other words, a stone would have the same (mass) (weight) whether it is on Earth's surface or the Moon's surface. However, its (mass) (weight) depends on its location.

On the Moon's surface, where gravity is only about 1/16 of Earth's gravity, (mass) (weight) (both the mass and the weight) of the stone would be the same as on Earth.

While mass and weight are not the same, they are (directly proportional) (inversely proportional) to each other.

In the same location, twice the mass has (twice) (half) the weight.

The Standard International (SI) unit of mass is the (kilogram) [newton], and the SI unit of force is the (kilogram) (newton).

In the United States, it is common to measure the mass of something by measuring its gravitational pull to Earth, its weight. The common unit of weight in the United States is the (pound) (kilogram) (newton).

When I step on a scale, two forces act on it; a downward pull of gravity, and an upward support force. These equal and opposite forces effectively compress a spring inside the scale that is calibrated to show weight. When in equilibrium, my weight = mg.

Pull of gravity

Support Force

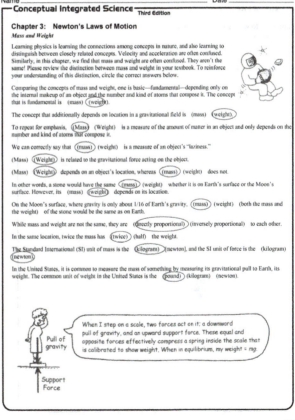

Chapter 3: Newton's Laws of Motion
Converting Mass to Weight

Objects with mass also have weight (although they can be weightless under special conditions). If you know the mass of something in **kilograms** and want its weight in **newtons**, at Earth's surface, you can take advantage of the formula that relates weight and mass:

$$\text{Weight} = \text{mass} \times \text{acceleration due to gravity}$$
$$W = mg.$$

This is in accord with Newton's Second Law, written as $F = ma$. When the force of gravity is the only force, the acceleration of any object of mass m will be g, the acceleration of free fall. Importantly, g acts as a proportionality constant, 9.8 N/kg, which is equivalent to 9.8 m/s^2.

Sample Question:

How much does a 1-kg bag of nails weigh on Earth?

$W = mg = (1 \text{ kg})(9.8 \text{ m/s}^2) = 9.8 \text{ m/s}^2 = 9.8 \text{ N}.$

or simply, $W = mg = (1 \text{ kg})(9.8 \text{ N/kg}) = 9.8 \text{ N}.$

From $F = ma$, we see that the unit of force equals the units [kg × m/s^2]. Can you see the units [m/s^2] ≈ [N/kg]?

Answer the following questions:

Felicia the ballet dancer has a mass of 45 kg.

1. What is Felicia's weight in newtons on Earth's surface? __441 N__

2. Given that 1 kilogram of mass corresponds to 2.2 pounds on Earth's surface, what is Felicia's weight in pounds on Earth? __99 LB__

3. What would be Felicia's mass on the surface of Jupiter? __45.0 kg__

4. What would be Felicia's weight on Jupiter's surface, where the acceleration due to gravity is 25.0 m/s^2?
__1125 N__

Different masses are hung on a spring scale calibrated in newtons. The force exerted by gravity on 1 kg = 9.8 N.

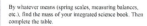

5. The force exerted by gravity on 5 kg = __49__ N.

6. The force exerted by gravity on __10__ kg = 98 N.

Make up your own mass and show the corresponding weight:

The force exerted by gravity on _____ kg = _____ N.

By whatever means (spring scales, measuring balances, etc.), find the mass of your integrated science book. Then complete the table.

OBJECT	MASS	WEIGHT
MELON	1 kg	9.8 N
APPLE	0.1 kg	1 N
BOOK		
A FRIEND	60 kg	588 N

Chapter 3: Newton's Laws of Motion
Bronco and Newton's Second Law

Bronco skydives and parachutes from a stationary helicopter. Various stages of fall are shown in positions *a* through *f*. Using Newton's Second Law

$$a = \frac{F_{NET}}{m} = \frac{W - R}{m}$$

find Bronco's acceleration at each position (answer in the blanks to the right). You need to know that Bronco's mass *m* is 100 kg so his weight is a constant 1000 N. Air resistance *R* varies with speed and cross-sectional area as shown.

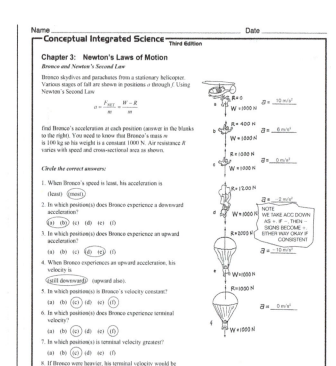

Circle the correct answers:

1. When Bronco's speed is least, his acceleration is
   (least) (most)

2. In which position(s) does Bronco experience a downward acceleration?
   (a) (b) (c) (d) (e) (f)

3. In which position(s) does Bronco experience an upward acceleration?
   (a) (b) (c) (d) (e) (f)

4. When Bronco experiences an upward acceleration, his velocity is
   (still downward) (upward also).

5. In which position(s) is Bronco's velocity constant?
   (a) (b) (c) (d) (e) (f)

6. In which position(s) does Bronco experience terminal velocity?
   (a) (b) (c) (d) (e) (f)

7. In which position(s) is terminal velocity greatest?
   (a) (b) (c) (d) (e) (f)

8. If Bronco were heavier, his terminal velocity would be
   (greater) (less) (the same).

Chapter 3: Newton's Laws of Motion
Newton's Third Law

Your thumb and finger pull on each other when you stretch a rubber band between them. This pair of forces, thumb on finger and finger on thumb, make up an action–reaction pair of forces, both of which are equal in magnitude and oppositely directed. Draw the reaction vector and state in words the reaction force for each of the examples **a** through **g**. Then make up your own example in **h**.

Thumb pulls finger
Finger pulls thumb

Foot hits ball
a BALL HITS FOOT

White ball strikes black ball
b BLACK BALL STRIKES WHITE BALL

Earth pulls on the Moon
c MOON PULLS ON EARTH

Tires push backward on road
d ROAD PUSHES FORWARD ON TIRES

Wings push air downward
e AIR PUSHES WINGS UPWARD

Fish pushes water backward
f WATER PUSHES FISH FORWARD

Helen touches Hyrum
g HYRUM TOUCHES HELEN

h OPEN: A ON B
   B ON A

YOU CAN'T TOUCH WITHOUT BEING TOUCHED— NEWTON'S THIRD LAW

Chapter 3: Newton's Laws of Motion
Nellie and Newton's Third Law

Nellie holds an apple weighing 1 newton at rest on the palm of her hand. *Circle the correct answers.*

1. To say the weight (W) of the apple is 1 N is to say that a downward gravitational force of 1 N is exerted on the apple by
   (Earth) (her hand).

2. Nellie's hand supports the apple with normal force N, which acts in a direction opposite to W. We can say N
   (equals W) (has the same magnitude as W)

3. Since the apple is at rest, the net force on the apple is
   (zero) (nonzero).

4. Since N is equal and opposite to W, we (can) (cannot) say that N and W constitute an action–reaction pair. The reason is that action and reaction (act on the same object) (act on different objects) and here we see N and W
   (both acting on the apple) (acting on different objects).

5. In accord with the rule "If ACTION is A acting on B, then REACTION is B acting on A." if we say action is Earth pulling down on the apple, reaction is
   (the apple pulling up on Earth) (N, Nellie's hand pushing up on the apple).

6. To repeat for emphasis, we see that N and W are equal and opposite to each other
   (and constitute an action–reaction pair) (but do *not* constitute an action–reaction pair).

To identify a pair of action–reaction forces in any situation, first identify the pair of interacting objects involved. Something is interacting with something else. In this case, the whole Earth is interacting (gravitationally) with the apple. So, Earth pulls downward on the apple (call it action), while the apple pulls upward on Earth (reaction).

Simply put, Earth pulls on apple (action), apple pulls on Earth (reaction).

Better put, apple and Earth pull on each other with equal and opposite forces that constitute a single interaction.

7. Another pair of forces is N [shown] and the downward force of the apple against Nellie's hand [not shown]. This pair of forces (is) (isn't) an action–reaction pair.

8. Suppose Nellie now pushes upward on the apple with the force of 2 N. The apple (is still in equilibrium) (accelerates upward) and compared with W, the magnitude of N is (the same) (twice)
   (not the same, and not twice).

9. Once the apple leaves Nellie's hand, N is (zero) (still twice the magnitude of W), and the net force on the apple is (zero) (only W) (still W – N, which is a negative force).

284

Conceptual Integrated Science — Third Edition

Chapter 3: Newton's Laws of Motion
Vectors and the Parallelogram Rule

1. When vectors **A** and **B** are at an angle to each other, they add to produce the resultant **C** by the *parallelogram rule*. Note that **C** is the diagonal of a parallelogram where **A** and **B** are adjacent sides. Resultant **C** is shown in the first two diagrams, *a* and *b*. Construct the resultant **C** in diagrams *c* and *d*. Note that in diagram *d* you form a rectangle (a special case of a parallelogram).

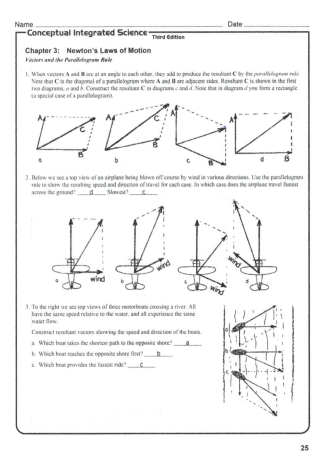

2. Below we see a top view of an airplane being blown off course by wind in various directions. Use the parallelogram rule to show the resulting speed and direction of travel for each case. In which case does the airplane travel fastest across the ground? __d__ Slowest? __c__

3. To the right we see top views of three motorboats crossing a river. All have the same speed relative to the water, and all experience the same water flow.

Construct resultant vectors showing the speed and direction of the boats.

a. Which boat takes the shortest path to the opposite shore? __a__

b. Which boat reaches the opposite shore first? __b__

c. Which boat provides the fastest ride? __c__

25

Conceptual Integrated Science — Third Edition

Vectors

Use the parallelogram rule to carefully construct the resultants for the eight pairs of vectors.

Carefully construct the vertical and horizontal components of the eight vectors.

26

Conceptual Integrated Science — Third Edition

Chapter 3: Newton's Laws of Motion
Force Vectors and the Parallelogram Rule

1. The heavy ball is supported in each case by two strands of rope. The tension in each strand is shown by the vectors. Use the parallelogram rule to find the resultant of each vector pair.

Note it's the angle, not the length of the rope, that affects tension!

a. Is your resultant vector the same for each case? __YES__

b. How do you think the resultant vector compares to the weight of the ball?
__SAME (BUT OPPOSITE DIRECTION)__

2. Now let's do the opposite of what we've done above. More often, we know the weight of the suspended object, but we don't know the rope tensions. In each case below, the weight of the ball is shown by the vector **W**. Each dashed vector represents the resultant of the pair of rope tensions. Note that each is equal and opposite to vector **W** (they must be; otherwise the ball wouldn't be at rest).

a. Construct parallelograms where the ropes define adjacent sides and the dashed vectors are the diagonals.

b. How do the relative lengths of the sides of each parallelogram compare to rope tensions?

c. Draw rope-tension vectors, clearly showing their relative magnitudes.

3. A lantern is suspended as shown. Draw vectors to show the relative tensions in ropes **A**, **B**, and **C**. Do you see a relationship between your vectors **A** + **B** and vector **C**? Between vectors **A** + **C** and vector **B**?

Yes; A + B = –C   A + C = –B

27

Conceptual Integrated Science — Third Edition

Force-Vector Diagrams

In each case, a rock is acted on by one or more forces. Draw an accurate vector diagram showing all forces acting on the rock, and no other forces. Use a ruler, and do it in pencil so you can correct mistakes. The first two are done as examples. Show by the parallelogram rule in 2 that the vector sum of **A** + **B** is equal and opposite to **W** (i.e., **A** + **B** = – **W**). Do the same for 3 and 4. Draw and label vectors for the weight and normal forces in 5 to 10, and for the appropriate forces in 11 and 12.

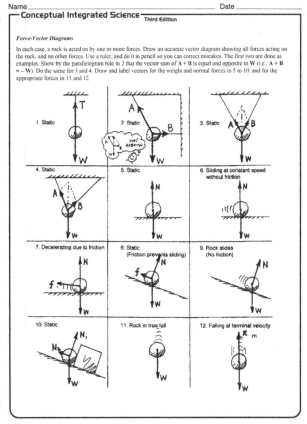

1. Static

2. Static

3. Static

4. Static

5. Static

6. Sliding at constant speed without friction

7. Decelerating due to friction

8. Static (Friction prevents sliding)

9. Rock slides (No friction)

10. Static

11. Rock in free fall

12. Falling at terminal velocity

28

285

Conceptual Integrated Science
Third Edition

Chapter 4: Momentum and Energy
Momentum

1. A moving car has momentum. If it moves twice as fast, its momentum is __TWICE__ as much.

2. Two cars, one twice as heavy as the other, move down a hill at the same speed. Compared with the lighter car, the momentum of the heavier car is __TWICE__ as much.

3. The recoil momentum of a gun that kicks is

   (more than)   (less than)   (the same as)

   the momentum of the gases and bullet it fires.

4. If a man firmly holds a gun when fired, then the momentum of the bullet and expelled gases is equal to the recoil momentum of the

   (gun alone)   (gun–man system)   (man alone).

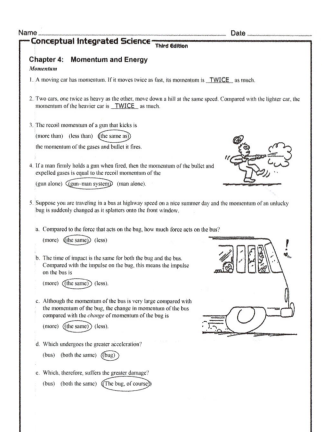

5. Suppose you are traveling in a bus at highway speed on a nice summer day and the momentum of an unlucky bug is suddenly changed as it splatters onto the front window.

   a. Compared to the force that acts on the bug, how much force acts on the bus?

   (more)   (the same)   (less)

   b. The time of impact is the same for both the bug and the bus. Compared with the impulse on the bug, this means the impulse on the bus is

   (more)   (the same)   (less).

   c. Although the momentum of the bus is very large compared with the momentum of the bug, the change in momentum of the bus compared with the *change* of momentum of the bug is

   (more)   (the same)   (less).

   d. Which undergoes the greater acceleration?

   (bus)   (both the same)   (bug)

   e. Which, therefore, suffers the greater damage?

   (bus)   (both the same)   (The bug, of course!)

Conceptual Integrated Science
Third Edition

Chapter 4: Momentum and Energy
Systems

Momentum conservation (and Newton's Third Law) applies to *systems* of bodies. Here we identify some systems.

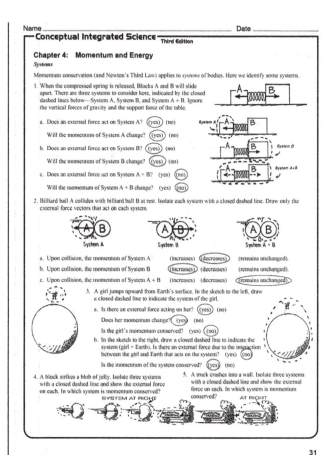

1. When the compressed spring is released, Blocks A and B will slide apart. There are three systems to consider here, indicated by the closed dashed lines below—System A, System B, and System A + B. Ignore the vertical forces of gravity and the support force of the table.

   a. Does an external force act on System A?   (yes)   (no)

   Will the momentum of System A change?   (yes)   (no)

   b. Does an external force act on System B?   (yes)   (no)

   Will the momentum of System B change?   (yes)   (no)

   c. Does an external force act on System A + B?   (yes)   (no)

   Will the momentum of System A + B change?   (yes)   (no)

2. Billiard ball A collides with billiard ball B at rest. Isolate each system with a closed dashed line. Draw only the external force vectors that act on each system.

   a. Upon collision, the momentum of System A   (increases)   (decreases)   (remains unchanged).

   b. Upon collision, the momentum of System B   (increases)   (decreases)   (remains unchanged).

   c. Upon collision, the momentum of System A + B   (increases)   (decreases)   (remains unchanged).

   3. A girl jumps upward from Earth's surface. In the sketch to the left, draw a closed dashed line to indicate the system of the girl.

      a. Is there an external force acting on her?   (yes)   (no)

      Does her momentum change?   (yes)   (no)

      Is the girl's momentum conserved?   (yes)   (no)

      b. In the sketch to the right, draw a closed dashed line to indicate the system (girl + Earth). Is there an external force due to the interaction between the girl and Earth that acts on the system?   (yes)   (no)

      Is the momentum of the system conserved?   (yes)   (no)

4. A block strikes a blob of jelly. Isolate three systems with a closed dashed line and show the external force on each. In which system is momentum conserved?   SYSTEM AT RIGHT

5. A truck crashes into a wall. Isolate three systems with a closed dashed line and show the external force on each. In which system is momentum conserved?   AT RIGHT

Conceptual Integrated Science
Third Edition

Chapter 4: Momentum and Energy
Impulse–Momentum

Bronco Brown wants to put $Ft = \Delta mv$ to the test and try bungee jumping. Bronco leaps from a high cliff and experiences free fall for 3 seconds. Then the bungee cord begins to stretch, reducing his speed to zero in 2 seconds. Fortunately, the cord stretches to its maximum length just short of the ground below.

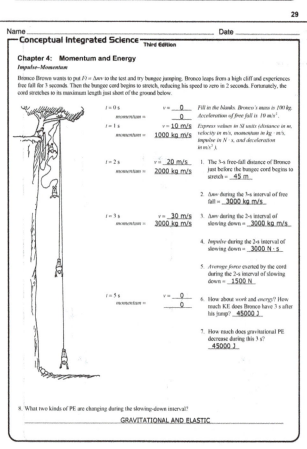

$t = 0$ s $v =$ __0__
momentum = __0__

$t = 1$ s $v =$ __10 m/s__
momentum = __1000 kg m/s__

$t = 2$ s $v =$ __20 m/s__
momentum = __2000 kg m/s__

$t = 3$ s $v =$ __30 m/s__
momentum = __3000 kg m/s__

$t = 5$ s $v =$ __0__
momentum = __0__

Fill in the blanks. Bronco's mass is 100 kg. Acceleration of free fall is 10 m/s². Express values in SI units (distance in m, velocity in m/s, momentum in kg · m/s, impulse in N · s, and deceleration in m/s²).

1. The 3-s free-fall distance of Bronco just before the bungee cord begins to stretch = __45 m__.

2. Δmv during the 3-s interval of free fall = __3000 kg m/s__.

3. Δmv during the 2-s interval of slowing down = __3000 kg m/s__.

4. *Impulse* during the 2-s interval of slowing down = __3000 N · s__.

5. *Average force* exerted by the cord during the 2-s interval of slowing down = __1500 N__.

6. How about *work* and *energy*? How much KE does Bronco have 3 s after his jump? __45000 J__

7. How much does gravitational PE decrease during this 3 s? __45000 J__

8. What two kinds of PE are changing during the slowing-down interval?
   __GRAVITATIONAL AND ELASTIC__

Conceptual Integrated Science
Third Edition

Chapter 4: Momentum and Energy
Conservation of Momentum

Granny whizzes around the rink and is suddenly confronted with Ambrose at rest directly in her path. Rather than knock him over, she picks him up and continues in motion without "braking." Consider both Granny and Ambrose as two parts of one system. Since no outside forces act on the system, the momentum of the system before collision equals the momentum of the system after collision.

   a. Complete the before-collision data in the table below.

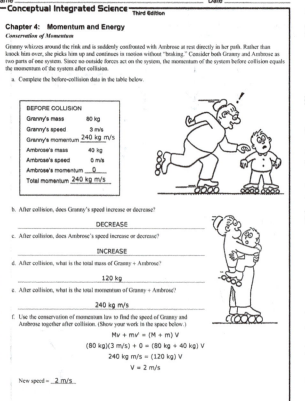

BEFORE COLLISION	
Granny's mass	80 kg
Granny's speed	3 m/s
Granny's momentum	240 kg m/s
Ambrose's mass	40 kg
Ambrose's speed	0 m/s
Ambrose's momentum	0
Total momentum	240 kg m/s

   b. After collision, does Granny's speed increase or decrease?

   __DECREASE__

   c. After collision, does Ambrose's speed increase or decrease?

   __INCREASE__

   d. After collision, what is the total mass of Granny + Ambrose?

   __120 kg__

   e. After collision, what is the total momentum of Granny + Ambrose?

   __240 kg m/s__

   f. Use the conservation of momentum law to find the speed of Granny and Ambrose together after collision. (Show your work in the space below.)

$$Mv + mv' = (M + m)V$$
$$(80 \text{ kg})(3 \text{ m/s}) + 0 = (80 \text{ kg} + 40 \text{ kg})V$$
$$240 \text{ kg m/s} = (120 \text{ kg})V$$
$$V = 2 \text{ m/s}$$

New speed = __2 m/s__

Chapter 4: Momentum and Energy
Work and Energy

1. How much work (energy) is needed to lift an object that weighs 200 N to a height of 4 m?

_____ 800 J _____

2. How much power is needed to lift the 200-N object to a height of 4 m in 4 s?

_____ 200 W _____

3. What is the power output of an engine that does 60,000 J of work in 10 s?

_____ 6 kW _____

4. The block of ice weighs 500 newtons.

a. Neglecting friction, how much force is needed to push it up the incline?

_____ 250 N _____

b. How much work is required to push it up the incline compared with lifting the block vertically 3 m?

_____ SAME (250 × 6 = 500 × 3) _____

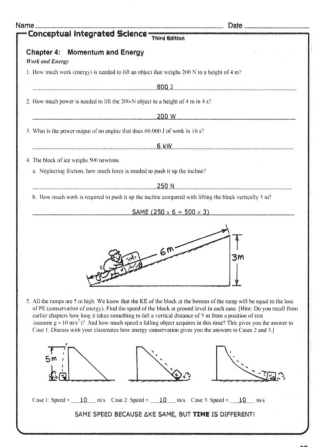

5. All the ramps are 5 m high. We know that the KE of the block at the bottom of the ramp will be equal to the loss of PE (conservation of energy). Find the speed of the block at ground level in each case. [Hint: Do you recall from earlier chapters how long it takes something to fall a vertical distance of 5 m from a position of rest (assume g = 10 m/s²)? And how much speed a falling object acquires in this time? This gives you the answer to Case 1. Discuss with your classmates how energy conservation gives you the answers to Cases 2 and 3.]

Case 1: Speed = 10 m/s   Case 2: Speed = 10 m/s   Case 3: Speed = 10 m/s

SAME SPEED BECAUSE ΔKE SAME, BUT **TIME** IS DIFFERENT!

Work and Energy—continued

6. Which block gets to the bottom of the incline first? Assume there is no friction. (Be careful!) Explain your answer.

BLOCK A GETS TO THE BOTTOM FIRST. IT HAS MORE ACCELERATION (STEEPER) AND LESS SLIDING DISTANCE—(HOWEVER, BOTH HAVE SAME **SPEED** AT BOTTOM—BUT WE'RE ASKED FOR **TIME**)

7. The KE and PE of a block freely sliding down a ramp are shown in only one place in the sketch. Fill in the missing information.

PE = 75 J   KE = 0
PE = 50 J   KE = 25 J
PE = 25 J   KE = 50 J
PE = 0   KE = 75 J

8. A big metal bead slides due to gravity along an upright friction-free wire. It starts from rest at the top of the wire as shown in the sketch. How fast is it traveling as it passes

Point B? 10 m/s

Point D? 10 m/s

Point E? 10 m/s

At what point does it have the maximum speed? C

9. Rows of wind-powered generators are used in various windy locations to generate electric power. Does the power generated affect the speed of the wind? Would locations behind the "windmills" be windier if they weren't there? Discuss this in terms of energy conservation with your classmates.

YES! BY CONS OF ENERGY, ENERGY GAINED BY WINDMILLS IS TAKEN FROM KE OF WIND—SO WIND MUST SLOW DOWN. LOCATIONS BEHIND WOULD BE A BIT WINDIER WITHOUT THE WINDMILLS!

THINK ENERGY CONVSERVATION!

Chapter 4: Momentum and Energy
Conservation of Energy

Fill in the blanks for the six systems shown:

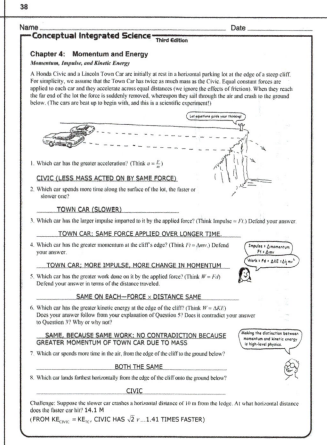

PE = 15000 J   KE = 0

$v = 30$ km/h   KE = 10⁶ J
$v = 60$ km/h   KE = 4 × 10⁶ J
$v = 90$ km/h   KE = 9 × 10⁶ J

PE = 11250 J   KE = 3750 J

PE = 30 J
PE = 30 J
PE = 30 J
PE = 20 J
PE = 0 J
KE = 30 J

PE = 7500 J   KE = 7500 J

PE = 3750 J   KE = 11250 J

PE = 10⁴ J

WORK DONE = 10⁴ J

PE = 50 J   KE = 0

PE = 0 J   KE = 15000 J

PE = 25 J   KE = 25 J

PE = 0   KE = 50 J

PE = 10 J   KE = 0
PE = 2 J   KE = 8 J
PE = 0   KE = 10 J
PE = 10 J   KE = 0

Chapter 4: Momentum and Energy
Momentum, Impulse, and Kinetic Energy

A Honda Civic and a Lincoln Town Car are initially at rest in a horizontal parking lot at the edge of a steep cliff. For simplicity, we assume that the Town Car has twice as much mass as the Civic. Equal constant forces are applied to each car and they accelerate across equal distances (we ignore the effects of friction). When they reach the far end of the lot the force is suddenly removed, whereupon they sail through the air and crash to the ground below. (The cars are beat up to begin with, and this is a scientific experiment!)

Let equations guide your thinking!

1. Which car has the greater acceleration? (Think $a = \frac{F}{m}$)

CIVIC (LESS MASS ACTED ON BY SAME FORCE)

2. Which car spends more time along the surface of the lot, the faster or slower one?

TOWN CAR (SLOWER)

3. Which car has the larger impulse imparted to it by the applied force? (Think Impulse = Ft.) Defend your answer.

TOWN CAR; SAME FORCE APPLIED OVER LONGER TIME.

4. Which car has the greater momentum at the cliff's edge? (Think Ft = Δmv.) Defend your answer.

TOWN CAR; MORE IMPULSE, MORE CHANGE IN MOMENTUM

Impulse = Δ momentum
Ft = Δmv
Work = Fd = ΔKE = Δ½mv²

5. Which car has the greater work done on it by the applied force? (Think W = Fd) Defend your answer in terms of the distance traveled.

SAME ON EACH—FORCE × DISTANCE SAME

6. Which car has the greater kinetic energy at the edge of the cliff? (Think W = ΔKE) Does your answer follow from your explanation of Question 5? Does it contradict your answer to Question 3? Why or why not?

SAME, BECAUSE SAME WORK; NO CONTRADICTION BECAUSE GREATER MOMENTUM OF TOWN CAR DUE TO MASS

Making the distinction between momentum and kinetic energy is high-level physics.

7. Which car spends more time in the air, from the edge of the cliff to the ground below?

BOTH THE SAME

8. Which car lands farthest horizontally from the edge of the cliff onto the ground below?

CIVIC

Challenge: Suppose the slower car crashes a horizontal distance of 10 m from the ledge. At what horizontal distance does the faster car hit? 14.1 M

(FROM KE$_{CIVIC}$ = KE$_{TC}$, CIVIC HAS $\sqrt{2}$ v...1.41 TIMES FASTER)

Conceptual Integrated Science *Third Edition*

Chapter 5: Gravity
The Inverse-Square Law—Weight

1. Paint spray travels radially away from the nozzle of the can in straight lines. Like gravity, the strength (intensity) of the spray obeys an inverse-square law. Complete the diagram by filling in the blank spaces.

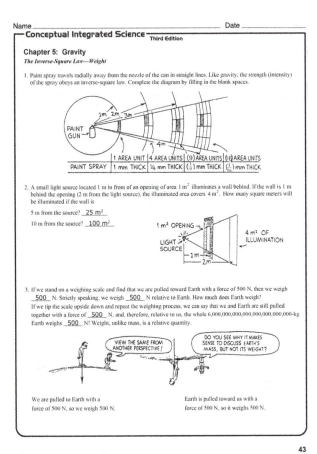

	1 AREA UNIT	4 AREA UNITS	(9) AREA UNITS	(16) AREA UNITS
PAINT SPRAY	1 mm THICK	¼ mm THICK	(1/9) mm THICK	(1/16) mm THICK

2. A small light source located 1 m in front of an opening of area 1 m² illuminates a wall behind. If the wall is 1 m behind the opening (2 m from the light source), the illuminated area covers 4 m². How many square meters will be illuminated if the wall is

5 m from the source? __25 m²__

10 m from the source? __100 m²__

3. If we stand on a weighing scale and find that we are pulled toward Earth with a force of 500 N, then we weigh __500__ N. Strictly speaking, we weigh __500__ N relative to Earth. How much does Earth weigh?
If we tip the scale upside down and repeat the weighing process, we can say that we and Earth are still pulled together with a force of __500__ N, and, therefore, relative to us, the whole 6,000,000,000,000,000,000,000,000-kg Earth weighs __500__ N! Weight, unlike mass, is a relative quantity.

VIEW THE SAME FROM ANOTHER PERSPECTIVE!

DO YOU SEE WHY IT MAKES SENSE TO DISCUSS EARTH'S MASS, BUT NOT ITS WEIGHT?

We are pulled to Earth with a force of 500 N, so we weigh 500 N.

Earth is pulled toward us with a force of 500 N, so it weighs 500 N.

Conceptual Integrated Science *Third Edition*

Chapter 5: Gravity
Ocean Tides

1. Consider two equal-mass blobs of water, A and B, initially at rest in the Moon's gravitational field. The vector shows the gravitational force of the Moon on A.

    a. Draw a force vector on B due to the Moon's gravity.

    b. Is the force on B more or less than the force on A? __LESS__

    c. Why? __FARTHER AWAY__

    d. The blobs accelerate toward the Moon. Which has the greater acceleration? (A) (B)

    e. Because of the different accelerations, with time

    (A gets farther ahead of B) (A and B gain identical speeds) and the distance between A and B

    (increases) (stays the same) (decreases).

    f. If A and B were connected by a rubber band, with time the rubber band would

    (stretch) (not stretch).

    g. This (stretching) (nonstretching) is due to the (difference) (nondifference) in the Moon's gravitational pulls.

    h. The two blobs will eventually crash into the Moon. To orbit around the Moon instead of crashing into it, the blobs should move (away from the Moon) (tangentially). Then their accelerations will consist of changes in (speed) (direction).

2. Now consider the same two blobs located on opposite sides of Earth.

    a. Because of differences in the Moon's pull on the blobs, they tend to

    (spread away from each other) (approach each other). This produces ocean tides!

    b. If Earth and the Moon were closer, gravitational force between them would be

    (more) (the same) (less), and the difference in gravitational forces on the near and far parts of the ocean

    would be (more) (the same) (less).

    c. Because Earth's orbit about the Sun is slightly elliptical, Earth and the Sun are closer in December than in June. Taking the Sun's tidal force into account, on a world average, ocean tides are greater in

    (December) (June) (no difference).

Conceptual Integrated Science *Third Edition*

Chapter 5: Gravity
Projectile Motion

5 m
20 m
45 m
80 m

1. Above left: Use the scale 1 cm:5 m and draw the positions of the dropped ball at 1-second intervals. Neglect air drag and assume *g* = 10 m/s². Estimate the number of seconds the ball is in the air.

    __4__ seconds.

2. Above right: The four positions of the thrown ball with *no gravity* are at 1-second intervals. At 1 cm:5 m, carefully draw the positions of the ball *with* gravity. Neglect air drag and assume *g* = 10 m/s². Connect your positions with a smooth curve to show the path of the ball. How is the motion in the vertical direction affected by motion in the horizontal direction?

    VERTICAL MOTION AFFECTED BY GRAVITY—HORIZONTAL MOTION DOESN'T AFFECT VERTICAL MOTION

Conceptual Integrated Science *Third Edition*

Projectile Motion—continued

5 m
20 m
45 m
80 m

3. This time the ball is thrown downward. Use the same scale 1 cm:5 m and carefully draw the positions of the ball as it falls beneath the dashed line. Connect your positions with a smooth curve. Estimate the number of seconds the ball remains in the air. __3.5__ seconds

4. Suppose you are an accident investigator and are asked to figure out whether the car was speeding before it crashed through the rail of the bridge and into the mudbank as shown. The speed limit on the bridge is 55 mph = 24 m/s. What is your conclusion?

    CAR COVERS 24 M IN 1 SEC (5 M DROP!), SO IT'S GOING 24 M/S AFTER CRASHING THROUGH RAIL. SO IT MUST HAVE BEEN GOING FASTER **BEFORE** HITTING RAIL. SO DRIVER WAS SPEEDING!

Chapter 5: Gravity
Tossed-Ball Vectors

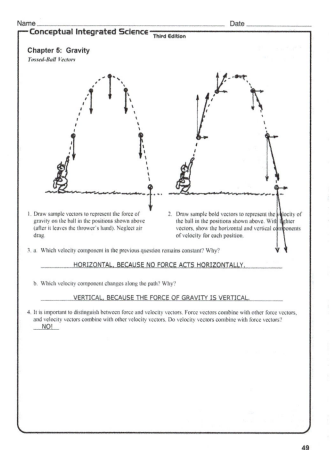

1. Draw sample vectors to represent the force of gravity on the ball in the positions shown above (after it leaves the thrower's hand). Neglect air drag.

2. Draw sample bold vectors to represent the velocity of the ball in the positions shown above. With lighter vectors, show the horizontal and vertical components of velocity for each position.

3. a. Which velocity component in the previous question remains constant? Why?

   HORIZONTAL, BECAUSE NO FORCE ACTS HORIZONTALLY.

   b. Which velocity component changes along the path? Why?

   VERTICAL, BECAUSE THE FORCE OF GRAVITY IS VERTICAL.

4. It is important to distinguish between force and velocity vectors. Force vectors combine with other force vectors, and velocity vectors combine with other velocity vectors. Do velocity vectors combine with force vectors?

   NO!

Tossed-Ball Vectors—continued

A ball tossed upward has initial velocity components 30 m/s vertical and 5 m/s horizontal. The position of the ball is shown at 1-second intervals. Air resistance is negligible, and $g = 10 \text{ m/s}^2$. Fill in the boxes, writing in the values of velocity *components* ascending, and your calculated *resultant velocities* descending.

10 m/s   5 m/s

5 m/s

20 m/s

11.2 m/s

Use the geometry theorem
$c^2 = a^2 + b^2$
to find the resultant velocities.

5 m/s

30 m/s

More specifically,
$v = \sqrt{v_x^2 + v_y^2}$

20.6 m/s

5 m/s

30.4 m/s

Chapter 5: Gravity
Circular and Elliptical Orbits

I. Circular Orbits

1. Figure 1 shows "Newton's Mountain," so high that its top is above the drag of the atmosphere. The cannonball is fired and hits the ground as shown.

   a. Draw the path the cannonball might take if it were fired a little bit faster.

   b. Repeat for a still greater speed, but still less than 8 km/s.

   c. Draw the orbital path it would take if its speed were 8 km/s.

   d. What is the shape of the 8-km/s curve?

   CIRCLE

   e. What would be the shape of the orbital path if the cannonball were fired at a speed of about 9 km/s?

   ELLIPSE

Figure 1

2. Figure 2 shows a satellite in circular orbit.

   a. At each of the four positions draw a vector that represents the gravitational *force* exerted on the satellite.

   b. Label the force vectors *F*.

   c. Draw at each position a vector to represent the *velocity* of the satellite at that position and label it *V*.

   d. Are all four *F* vectors the same length? Why or why not?

   YES; SAME DISTANCE, SAME FORCE

   Figure 2

   e. Are all four *V* vectors the same length? Why or why not?

   YES—IN CIRCULAR ORBIT F ⊥ v SO NO COMPONENT OF F ALONG v

   f. What is the angle between your *F* and *V* vectors? 90°

   g. Is there any component of *F* along *V*? NO (F⊥v)

   h. What does this tell you about the work the force of gravity does on the satellite?

   NO WORK, BECAUSE NO COMPONENT OF F ALONG PATH

   i. Does the KE of the satellite in Figure 2 remain constant, or does it vary? CONSTANT

   j. Does the PE of the satellite remain constant, or does it vary? CONSTANT

Circular and Elliptical Orbits—continued

II. Elliptical Orbits

3. Figure 3 shows a satellite in elliptical orbit.

   a. Repeat the procedure you used for the circular orbit, drawing vectors *F* and *V* for each position, including proper labeling. Show equal magnitudes with equal lengths, and greater magnitudes with greater lengths, but don't bother making the scale accurate.

   b. Are your vectors *F* all the same magnitude? Why or why not?

   NO, FORCE DECREASES WHEN DISTANCE FROM EARTH INCREASES.

   c. Are your vectors *V* all the same magnitude? Why or why not?

   NO, WHEN KE DECREASES, SPEED DECREASES. WHEN KE INCREASES (CLOSER TO EARTH) SPEED INCREASES.

   d. Is the angle between vectors *F* and *V* everywhere the same, or does it vary?

   IT VARIES.

   e. Are there places where there is a component of *F* along *V*?

   YES (EVERYWHERE EXCEPT AT THE APOGEE AND PERIGEE).

   f. Is work done on the satellite when there is a component of *F* along and in the same direction of *V*, and if so, does this increase or decrease the KE of the satellite?

   YES; THIS INCREASES KE OF SATELLITE.

   g. When there is a component of *F* along and opposite to the direction of *V*, does this increase or decrease the KE of the satellite?

   THIS DECREASES KE OF SATELLITE.

   Figure 3

   h. What can you say about the sum KE + PE along the orbit?

   CONSTANT (IN ACCORD WITH CONSERVATION OF ENERGY).

Be very, very careful when placing both velocity and force vectors on the same diagram. Not a good practice, for one may construct the resultant of the vectors—ouch!

Conceptual Integrated Science *Third Edition*

Chapter 5: Gravity
Mechanics Overview

1. The sketch shows the elliptical path described by a satellite about Earth. In which of the marked positions, A–D. (put S for "same everywhere") does the satellite experience the maximum

a. gravitational force? __A__

b. speed? __A__

c. velocity? __A__

d. momentum? __A__

e. kinetic energy? __A__

f. gravitational potential energy? __C__

g. total energy (KE + PE)? __S__

h. acceleration? __A__

$$a = \frac{F}{m}$$

2. Answer the above questions for a satellite in circular orbit.

a. __S__ b. __S__ c. __S__ d. __S__ e. __S__ f. __S__ g. __S__ h. __S__

3. In which position(s) is there momentarily no work done on the satellite by the force of gravity? Why?

A AND C, BECAUSE NO FORCE COMPONENTS ALONG PATH

4. Work changes energy. Let the equation for work, $W = Fd$, guide your thinking on these questions. Defend your answers in terms of $W = Fd$.

a. In which position will a several-minutes thrust of rocket engines do the most work on the satellite and give it the greatest change in kinetic energy?

A, BECAUSE d GREATEST DURING THRUST—F × d IS MORE WORK

b. In which position will a several-minutes thrust of rocket engines do the most work on the *exhaust gases* and give the *exhaust gases* the greatest change in kinetic energy?

C, WHERE THE ROCKET IS SLOWEST.

c. In which position will a several-minutes thrust of rocket engines give the satellite the least boost in kinetic energy?

C, BECAUSE RELATIVE TO PLANET, MOST ENERGY IS GIVEN TO THE EXHAUST GASES.

Conceptual Integrated Science *Third Edition*

Chapter 6: Heat
Temperature Mix

1. You apply heat to 1 L of water and raise its temperature by 10°C. If you add the same quantity of heat to 2 L of water, how much will the temperature rise? To 3 L of water?

Record your answers on the blanks in the drawing at the right. (Hint: Heat transferred is directly proportional to its temperature change, $Q = mc\Delta T$.)

ΔT = 10°C   ΔT = 5°C   ΔT = 3.3°C
1L   2L   3L

2. A large bucket contains 1 L of 20°C water.

a. What will be the temperature of the mixture when 1 L of 20°C water is added?

STILL 20°C

b. What will be the temperature of the mixture when 1 L of 40°C water is added?

30°C

c. If 2 L of 40°C water were added, would the temperature of the mixture be greater or less than 30°C?

GREATER

$$2 (40°C - T) = 1(T - 20°C)$$
$$T = 33.3°C$$

3. A red-hot iron kilogram mass is put into 1 L of cool water. Mark each of the following statements true (T) or false (F). (Ignore heat transfer to the container.)

a. The increase in the water temperature is equal to the decrease in the iron's temperature. NOTE DISTINCTION!

F

b. The quantity of heat gained by the water is equal to the quantity of heat lost by the iron.

T

c. The iron and the water will both reach the same temperature. THERMAL EQUILIBRIUM.

T

d. The final temperature of the iron and water is about halfway between the initial temperatures of each.

F

4. *True or False:* When Queen Elizabeth throws the last sip of her tea over Queen Mary's rail, the ocean gets a little warmer. T (UNLESS IT WAS ICE TEA!)

Conceptual Integrated Science *Third Edition*

Chapter 6: Heat
Absolute Zero

A mass of air is contained so that the volume can change but the pressure remains constant. Table 1 shows air volumes at various temperatures when the air is heated slowly.

1. Plot the data in Table 1 on the graph and connect the points.

TABLE 1

TEMP. (°C)	VOLUME (mL)
0	50
25	55
50	60
75	65
100	70

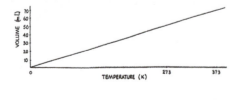

2. The graph shows how the volume of air varies with temperature at constant pressure. The straightness of the line means that the air expands uniformly with temperature. From your graph, you can predict what will happen to the volume of air when it is cooled.

Extrapolate (extend) the straight line of your graph to find the temperature at which the volume of the air would become zero. Mark this point on your graph. Estimate this temperature: __-273°C__

3. Although air would liquify before cooling to this temperature, the procedure suggests that there is a lower limit to how cold something can be. This is the absolute zero of temperature.

Careful experiments show that absolute zero is __-273__ °C.

4. Scientists measure temperature in *kelvins* instead of degrees Celsius, where the absolute zero of temperature is 0 kelvins. If you relabeled the temperature axis on the graph in Question 1 so that it shows temperature in kelvins, would your graph look like the one below? __YES__

Conceptual Integrated Science *Third Edition*

Chapter 6: Heat
Thermal Expansion

1. Steel expands by about 1 part in 100,000 for each 1°C increase in temperature.

$$\Delta l = \frac{1}{10^5} l_o \Delta T$$

a. How much longer will a piece of steel 1000 mm long (1 meter) be when its temperature is increased by 10°C? __0.1 mm__ $\Delta l = \frac{1}{10^5} l_o \Delta T = \frac{10^3}{10^5} 10 = \frac{10^4}{10^5} = 10^{-1} = 0.1$ mm

b. How much longer will a piece of steel 1000 m long (1 kilometer) be when its temperature is increased by 10°C? __0.1 m = 10 cm__

c. You place yourself between a wall and the end of a 1-m steel rod when the opposite end is securely fastened as shown. No harm comes to you if the temperature of the rod is increased a few degrees. Discuss the consequences of doing this with a rod many meters long.

Δl IS SMALL FOR SMALL l_o, BUT CAN BE FATALLY LARGE (YOUR BODY WIDTH!) FOR LARGE l_o.

2. The Eiffel Tower in Paris is 298 meters high. On a cold winter night, it is shorter than on a hot summer day. What is its change in height for a 30°C temperature difference?

$$\Delta l = \frac{298}{10^5} \cdot 30 = 0.09 \text{ m} = 9 \text{ cm}$$

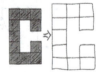

3. Consider a gap in a piece of metal. Does the gap become wider or narrower when the metal is heated? (Consider the piece of metal made up of 11 blocks—if the blocks are individually heated, each is slightly larger. Make a sketch of them, slightly enlarged, beside the sketch shown.)

GAP IS WIDER (AS MUCH IF IT WERE ALL METAL)

4. The equatorial radius of Earth is about 6370 km. Consider a 40,000-km long steel pipe that forms a giant ring that fits snugly around Earth's equator. Suppose people all along its length breathe on it so as to raise its temperature by 1°C. The pipe gets longer. It is also no longer snug. How high does it stand above the ground? (Hint: Concentrate on the radial distance.)

$$\Delta r = \frac{6370}{10^5} \text{ km} \cdot 10 = 0.637 \text{ km } 63.7 \text{ mi}$$ WOW!

Thermal Expansion—continued

5. A weight hangs above the floor from the copper wire. When a candle is moved along the wire and heats it, what happens to the height of the weight above the floor? Why?

HEIGHT DECREASES AS WIRE LENGTHENS

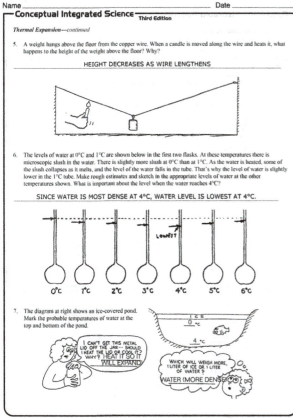

6. The levels of water at 0°C and 1°C are shown below in the first two flasks. At these temperatures there is microscopic slush in the water. There is slightly more slush at 0°C than at 1°C. As the water is heated, some of the slush collapses as it melts, and the level of the water falls in the tube. That's why the level of water is slightly lower in the 1°C tube. Make rough estimates and sketch in the appropriate levels of water at the other temperatures shown. What is important about the level when the water reaches 4°C?

SINCE WATER IS MOST DENSE AT 4°C, WATER LEVEL IS LOWEST AT 4°C.

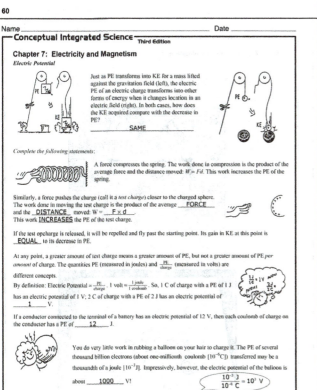

7. The diagram at right shows an ice-covered pond. Mark the probable temperatures of water at the top and bottom of the pond.

WATER (MORE DENSE)

Chapter 6: Heat
Transmission of Heat

1. The tips of both brass rods are held in the gas flame. *Mark the following true (T) or false (F).*

   a. Heat is conducted only along Rod A. __F__

   b. Heat is conducted only along Rod B. __F__

   c. Heat is conducted equally along both Rod A and Rod B. __T__

   d. The idea that "heat rises" applies to heat transfer by *convection*, not by *conduction*. __T__

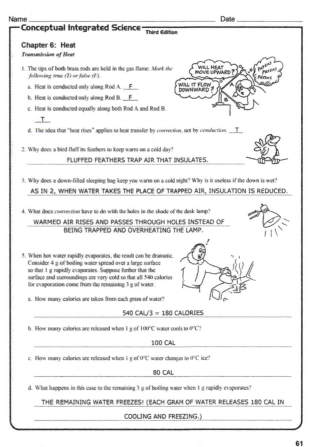

2. Why does a bird fluff its feathers to keep warm on a cold day?

FLUFFED FEATHERS TRAP AIR THAT INSULATES.

3. Why does a down-filled sleeping bag keep you warm on a cold night? Why is it useless if the down is wet?

AS IN 2, WHEN WATER TAKES THE PLACE OF TRAPPED AIR, INSULATION IS REDUCED.

4. What does *convection* have to do with the holes in the shade of the desk lamp?

WARMED AIR RISES AND PASSES THROUGH HOLES INSTEAD OF
BEING TRAPPED AND OVERHEATING THE LAMP.

5. When hot water rapidly evaporates, the result can be dramatic. Consider 4 g of boiling water spread over a large surface so that 1 g rapidly evaporates. Suppose further that the surface and surroundings are very cold so that all 540 calories for evaporation come from the remaining 3 g of water.

   a. How many calories are taken from each gram of water?

   540 CAL/3 = 180 CALORIES

   b. How many calories are released when 1 g of 100°C water cools to 0°C?

   100 CAL

   c. How many calories are released when 1 g of 0°C water changes to 0°C ice?

   80 CAL

   d. What happens in this case to the remaining 3 g of boiling water when 1 g rapidly evaporates?

   THE REMAINING WATER FREEZES! (EACH GRAM OF WATER RELEASES 180 CAL IN

   COOLING AND FREEZING.)

Chapter 7: Electricity and Magnetism
Electric Potential

Just as PE transforms into KE for a mass lifted against the gravitation field (left), the electric PE of an electric charge transforms into other forms of energy when it changes location in an electric field (right). In both cases, how does the KE acquired compare with the decrease in PE?

SAME

Complete the following statements:

A force compresses the spring. The work done in compression is the product of the average force and the distance moved: $W = Fd$. This work increases the PE of the spring.

Similarly, a force pushes the charge (call it a *test charge*) closer to the charged sphere. The work done in moving the test charge is the product of the average __FORCE__ and the __DISTANCE__ moved: $W = \underline{F \times d}$. This work __INCREASES__ the PE of the test charge.

If the test opcharge is released, it will be repelled and fly past the starting point. Its gain in KE at this point is __EQUAL__ to its decrease in PE.

At any point, a greater amount of test charge means a greater amount of PE, but not a greater amount of PE *per amount* of charge. The quantities PE (measured in joules) and $\frac{PE}{charge}$ (measured in volts) are different concepts.

By definition: Electric Potential = $\frac{PE}{charge}$. 1 volt = $\frac{1\ joule}{1\ coulomb}$. So, 1 C of charge with a PE of 1 J has an electric potential of __1__ V; 2 C of charge with a PE of 2 J has an electric potential of __1__ V.

If a conductor connected to the terminal of a battery has an electric potential of 12 V, then each coulomb of charge on the conductor has a PE of __12__ J.

You do very little work in rubbing a balloon on your hair to charge it. The PE of several thousand billion electrons (about one-millionth coulomb $[10^{-6}C]$) transferred may be a thousandth of a joule $[10^{-3}J]$. Impressively, however, the electric potential of the balloon is about __1000__ V!

$$\frac{10^{-3}\ J}{10^{-6}\ C} = 10^3\ V$$

Why is contact with a balloon charged to thousands of volts not as dangerous as contact with household 110 V?

HOUSEHOLD CURRENT TRANSFERS MANY COULOMBS AND MUCH ENERGY.

A BALLOON TRANSFERS VERY LITTLE OF BOTH.

Chapter 7: Electricity and Magnetism
Series Circuits

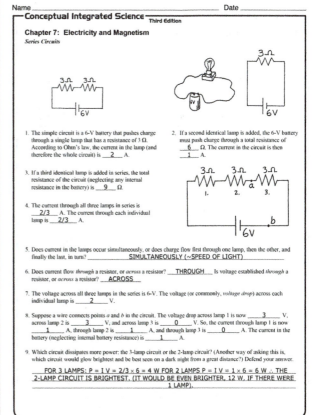

1. The simple circuit is a 6-V battery that pushes charge through a single lamp that has a resistance of 3 Ω. According to Ohm's law, the current in the lamp (and therefore the whole circuit) is __2__ A.

2. If a second identical lamp is added, the 6-V battery must push charge through a total resistance of __6__ Ω. The current in the circuit is then __1__ A.

3. If a third identical lamp is added in series, the total resistance of the circuit (neglecting any internal resistance in the battery) is __9__ Ω.

4. The current through all three lamps in series is __2/3__ A. The current through each individual lamp is __2/3__ A.

5. Does current in the lamps occur simultaneously, or does charge flow first through one lamp, then the other, and finally the last, in turn? __SIMULTANEOUSLY (~SPEED OF LIGHT)__

6. Does current flow *through* a resistor, or *across* a resistor? __THROUGH__ Is voltage established *through* a resistor, or *across* a resistor? __ACROSS__

7. The voltage across all three lamps in the series is 6-V. The voltage (or commonly, *voltage drop*) across each individual lamp is __2__ V.

8. Suppose a wire connects points *a* and *b* in the circuit. The voltage drop across lamp 1 is now __3__ V, across lamp 2 is __3__ V, and across lamp 3 is __0__ V. So, the current through lamp 1 is now __1__ A, through lamp 2 is __1__ A, and through lamp 3 is __0__ A. The current in the battery (neglecting internal battery resistance) is __1__ A.

9. Which circuit dissipates more power: the 3-lamp circuit or the 2-lamp circuit? (Another way of asking this is, which circuit would glow brightest and be best seen on a dark night from a great distance?) Defend your answer.

FOR 3 LAMPS: P = I V = 2/3 × 6 = 4 W FOR 2 LAMPS P = I V = 1 × 6 = 6 W ∴ THE
2-LAMP CIRCUIT IS BRIGHTEST. (IT WOULD BE EVEN BRIGHTER, 12 W, IF THERE WERE
1 LAMP).

Chapter 7: Electricity and Magnetism
Parallel Circuits

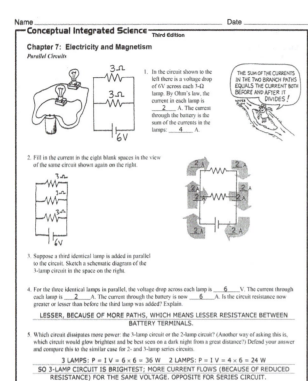

1. In the circuit shown to the left there is a voltage drop of 6V across each 3-Ω lamp. By Ohm's law, the current in each lamp is ___2___ A. The current through the battery is the sum of the currents in the lamps: ___4___ A.

THE SUM OF THE CURRENTS IN THE TWO BRANCH PATHS EQUALS THE CURRENT BOTH BEFORE AND AFTER IT DIVIDES!

2. Fill in the current in the eight blank spaces in the view of the same circuit shown again on the right.

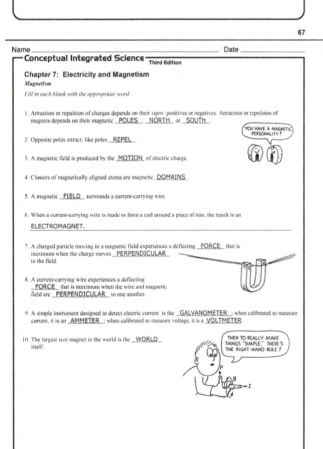

3. Suppose a third identical lamp is added in parallel to the circuit. Sketch a schematic diagram of the 3-lamp circuit in the space on the right.

4. For the three identical lamps in parallel, the voltage drop across each lamp is ___6___ V. The current through each lamp is ___2___ A. The current through the battery is now ___6___ A. Is the circuit resistance now greater or lesser than before the third lamp was added? Explain.

LESSER, BECAUSE OF MORE PATHS, WHICH MEANS LESSER RESISTANCE BETWEEN BATTERY TERMINALS.

5. Which circuit dissipates more power: the 3-lamp circuit or the 2-lamp circuit? (Another way of asking this is, which circuit would glow brightest and be best seen on a dark night from a great distance?) Defend your answer and compare this to the similar case for 2- and 3-lamp series circuits.

3 LAMPS: P = I V = 6 × 6 = 36 W   2 LAMPS: P = I V = 4 × 6 = 24 W
SO 3-LAMP CIRCUIT IS BRIGHTEST; MORE CURRENT FLOWS (BECAUSE OF REDUCED RESISTANCE) FOR THE SAME VOLTAGE. OPPOSITE FOR SERIES CIRCUIT.

Chapter 7: Electricity and Magnetism
Compound Circuits

The table beside circuit *a* below shows the current through each resistor, the voltage across each resistor, and the power dissipated as heat in each resistor. Find the similar correct values for circuits *b*, *c*, and *d*, and put your answers in the tables shown.

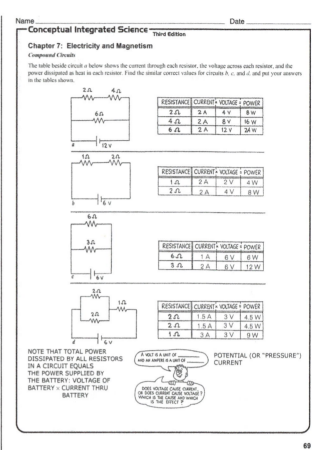

RESISTANCE	CURRENT ×	VOLTAGE =	POWER
2 Ω	2 A	4 V	8 W
4 Ω	2 A	8 V	16 W
6 Ω	2 A	12 V	24 W

RESISTANCE	CURRENT ×	VOLTAGE =	POWER
1 Ω	2 A	2 V	4 W
2 Ω	2 A	4 V	8 W

RESISTANCE	CURRENT ×	VOLTAGE =	POWER
6 Ω	1 A	6 V	6 W
3 Ω	2 A	6 V	12 W

RESISTANCE	CURRENT ×	VOLTAGE =	POWER
2 Ω	1.5 A	3 V	4.5 W
2 Ω	1.5 A	3 V	4.5 W
1 Ω	3 A	3 V	9 W

NOTE THAT TOTAL POWER DISSIPATED BY ALL RESISTORS IN A CIRCUIT EQUALS THE POWER SUPPLIED BY THE BATTERY: VOLTAGE OF BATTERY × CURRENT THRU BATTERY

A VOLT IS A UNIT OF _____ POTENTIAL (OR "PRESSURE")
AND AN AMPERE IS A UNIT OF _____ CURRENT

DOES VOLTAGE CAUSE CURRENT, OR DOES CURRENT CAUSE VOLTAGE? WHICH IS THE CAUSE AND WHICH IS THE EFFECT?

Chapter 7: Electricity and Magnetism
Magnetism

Fill in each blank with the appropriate word:

1. Attraction or repulsion of charges depends on their *signs*: positives or negatives. Attraction or repulsion of magnets depends on their magnetic __POLES__, __NORTH__ or __SOUTH__.

YOU HAVE A MAGNETIC PERSONALITY!

2. Opposite poles attract; like poles __REPEL__.

3. A magnetic field is produced by the __MOTION__ of electric charge.

4. Clusters of magnetically aligned atoms are magnetic __DOMAINS__.

5. A magnetic __FIELD__ surrounds a current-carrying wire.

6. When a current-carrying wire is made to form a coil around a piece of iron, the result is an __ELECTROMAGNET.__

7. A charged particle moving in a magnetic field experiences a deflecting __FORCE__ that is maximum when the charge moves __PERPENDICULAR__ to the field.

8. A current-carrying wire experiences a deflecting __FORCE__ that is maximum when the wire and magnetic field are __PERPENDICULAR__ to one another.

9. A simple instrument designed to detect electric current is the __GALVANOMETER__; when calibrated to measure current, it is an __AMMETER__; when calibrated to measure voltage, it is a __VOLTMETER__.

10. The largest size magnet in the world is the __WORLD__ itself.

THEN TO REALLY MAKE THINGS "SIMPLE," THERE'S THE RIGHT-HAND RULE!

Chapter 7: Electricity and Magnetism
Field Patterns

1. The illustration below is similar to Figure 7.32 in your textbook. Iron filings trace out patterns of magnetic field lines about a bar magnet. In the field are some magnetic compasses. The compass needle in only one compass is shown. Draw in the needles with proper orientation in the other compasses.

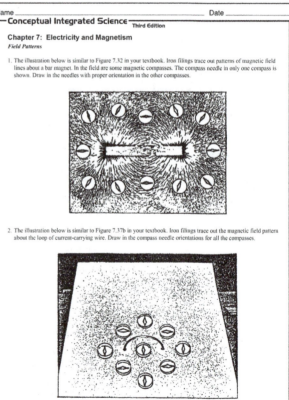

2. The illustration below is similar to Figure 7.37b in your textbook. Iron filings trace out the magnetic field pattern about the loop of current-carrying wire. Draw in the compass needle orientations for all the compasses.

292

Chapter 7: Electricity and Magnetism
Electromagnetism

1. Early investigators discovered that magnetism and electricity are
   (related) (independent of each other).
   Magnetism is produced by
   (batteries) (the motion of electric charges).

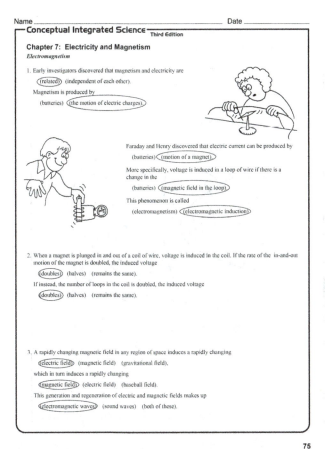

   Faraday and Henry discovered that electric current can be produced by
   (batteries) (motion of a magnet).
   More specifically, voltage is induced in a loop of wire if there is a change in the
   (batteries) (magnetic field in the loop).
   This phenomenon is called
   (electromagnetism) (electromagnetic induction).

2. When a magnet is plunged in and out of a coil of wire, voltage is induced in the coil. If the rate of the in-and-out motion of the magnet is doubled, the induced voltage
   (doubles) (halves) (remains the same).
   If instead, the number of loops in the coil is doubled, the induced voltage
   (doubles) (halves) (remains the same).

3. A rapidly changing magnetic field in any region of space induces a rapidly changing
   (electric field) (magnetic field) (gravitational field),
   which in turn induces a rapidly changing
   (magnetic field) (electric field) (baseball field).
   This generation and regeneration of electric and magnetic fields makes up
   (electromagnetic waves) (sound waves) (both of these).

Chapter 8: Waves—Sound and Light
Vibration and Wave Fundamentals

1. A sine curve that represents a transverse wave is drawn below. With a ruler, measure the wavelength and amplitude of the wave.

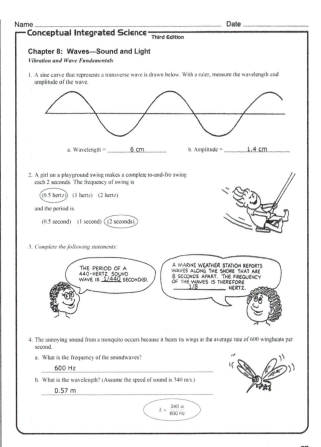

   a. Wavelength = ____6 cm____   b. Amplitude = ____1.4 cm____

2. A girl on a playground swing makes a complete to-and-fro swing each 2 seconds. The frequency of swing is
   (0.5 hertz) (1 hertz) (2 hertz)
   and the period is
   (0.5 second) (1 second) (2 seconds).

3. *Complete the following statements:*

   THE PERIOD OF A 440-HERTZ SOUND WAVE IS _1/440_ SECOND(S).

   A MARINE WEATHER STATION REPORTS WAVES ALONG THE SHORE THAT ARE 8 SECONDS APART. THE FREQUENCY OF THE WAVES IS THEREFORE _1/8_ HERTZ.

4. The annoying sound from a mosquito occurs because it beats its wings at the average rate of 600 wingbeats per second.
   a. What is the frequency of the soundwaves?
      ____600 Hz____
   b. What is the wavelength? (Assume the speed of sound is 340 m/s.)
      ____0.57 m____

   $$\lambda = \frac{340 \ a}{600 \ Hz}$$

Vibration and Wave Fundamentals—continued

5. A machine gun fires 10 rounds per second. The speed of the bullets is 300 m/s.

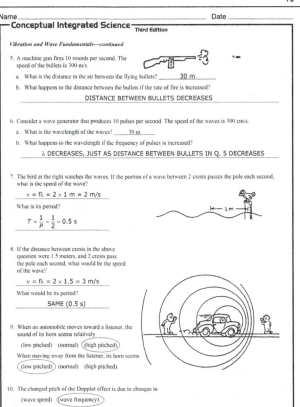

   a. What is the distance in the air between the flying bullets? ____30 m____
   b. What happens to the distance between the bullets if the rate of fire is increased?
      DISTANCE BETWEEN BULLETS DECREASES

6. Consider a wave generator that produces 10 pulses per second. The speed of the waves is 300 cm/s.
   a. What is the wavelength of the waves? ____30 cm____
   b. What happens to the wavelength if the frequency of pulses is increased?
      λ DECREASES, JUST AS DISTANCE BETWEEN BULLETS IN Q. 5 DECREASES

7. The bird at the right watches the waves. If the portion of a wave between 2 crests passes the pole each second, what is the speed of the wave?
   $v = f\lambda = 2 \times 1\ m = 2\ m/s$
   What is its period?
   $T = \frac{1}{P} = \frac{1}{2} = 0.5\ s$

8. If the distance between crests in the above question were 1.5 meters, and 2 crests pass the pole each second, what would be the speed of the wave?
   $v = f\lambda = 2 \times 1.5 = 3\ m/s$
   What would be its period?
   SAME (0.5 s)

9. When an automobile moves toward a listener, the sound of its horn seems relatively
   (low pitched) (normal) (high pitched).
   When moving away from the listener, its horn seems
   (low pitched) (normal) (high pitched).

10. The changed pitch of the Doppler effect is due to changes in
    (wave speed) (wave frequency).

Chapter 8: Waves—Sound and Light
Color

The sketch to the right shows the shadow of an instructor in front of a white screen in a dark room. The light source is red, so the screen looks red and the shadow looks black. Color the sketch, or label the colors with a pen or pencil.

A green lamp is added and makes a second shadow. The shadow cast by the red lamp is no longer black, but is illuminated by green light, so it is green. Color or mark it green. The shadow cast by the green lamp is not black because it is illuminated by the red lamp. Indicate its color. Do the same for the background, which receives a mixture of red and green light.

A blue lamp is added and three shadows appear. Indicate the appropriate colors of the shadows and the background.

The lamps are placed a bit closer together so the shadows overlap. Indicate the colors of all screen areas.

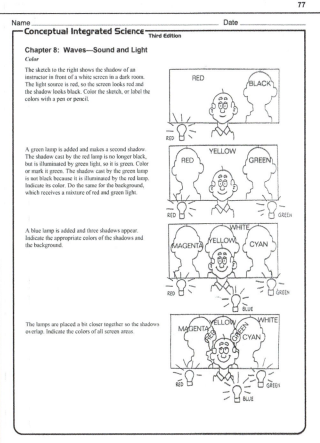

293

⌐Conceptual Integrated Science ⌐ *Third Edition*

Color—continued

If you have colored pencils or markers, have a go at these.

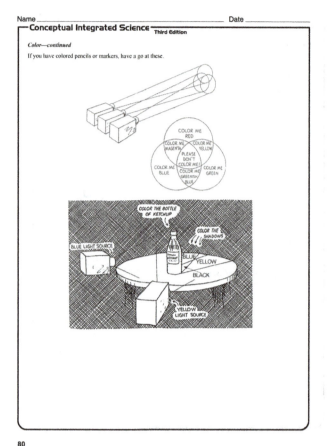

⌐Conceptual Integrated Science ⌐ *Third Edition*

Chapter 8: Waves—Sound and Light
Diffraction and Interference

Shown below are concentric solid and dashed circles, each different in radius by 1 cm. Consider the circular pattern of a top view of water waves, where the solid circles are crests and the dashed circles are troughs.

1. Draw another set of the same concentric circles with a compass. Choose any part of the paper for your center (except the present central point). Let the circles run off the edge of the paper.

2. Find where a dashed line crosses a solid line and draw a large dot at the intersection. Do this for ALL places where a solid and dashed line intersect.

3. With a wide felt marker, connect the dots with smooth lines. These *nodal lines* lie in regions where the waves have cancelled—where the crest of one wave overlaps the trough of another.

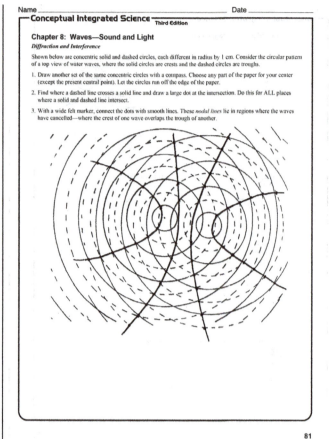

⌐Conceptual Integrated Science ⌐ *Third Edition*

Chapter 8: Waves—Sound and Light
Reflection

1. Light from a flashlight shines on a mirror and illuminates one of the cards. Draw the reflected beam to indicate the illuminated card.

2. A periscope has a pair of mirrors in it. Draw the light path from the object "O" to the eye of the observer.

3. The ray diagram below shows the extension of one of the reflected rays from the plane mirror. Complete the diagram by (1) carefully drawing the three other reflected rays and (2) extending them behind the mirror to locate the image of the flame. (Assume the candle and image are viewed by an observer on the left.)

⌐Conceptual Integrated Science ⌐ *Third Edition*

Reflection—continued

4. The ray diagram below shows the reflection of one of the rays that strikes the parabolic mirror. Notice that the law of reflection is observed, and the angle of incidence (from the normal, the dashed line) equals the angle of reflection (from the normal). Complete the diagram by drawing the reflected rays of the other three rays that are shown. (Do you see why parabolic mirrors are used in automobile headlights?)

5. A girl takes a photograph of the bridge as shown. Which of the two sketches below correctly shows the reflected view of the bridge? Defend your answer.

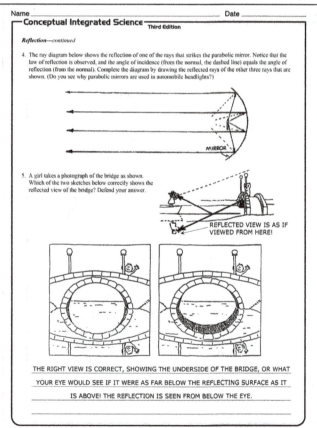

THE RIGHT VIEW IS CORRECT, SHOWING THE UNDERSIDE OF THE BRIDGE, OR WHAT

YOUR EYE WOULD SEE IF IT WERE AS FAR BELOW THE REFLECTING SURFACE AS IT

IS ABOVE! THE REFLECTION IS SEEN FROM BELOW THE EYE.

Conceptual Integrated Science *Third Edition*

Chapter 8: Waves—Sound and Light
Refraction—Part 1

1. A pair of toy cart wheels are rolled obliquely from a smooth surface onto two plots of grass—a rectangular plot as shown at the left, and a triangular plot as shown the right. The ground is on a slight incline, so that after slowing down in the grass, the wheels speed up again when emerging on the smooth surface. Finish each sketch and show some positions of the wheels inside the plots and on the other side. Clearly indicate their paths and directions of travel.

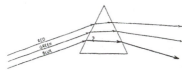

2. Red, green, and blue rays of light are incident upon a glass prism as shown. The average speed of red light in the glass is less than in air, so the red ray is refracted. When it emerges into the air it regains its original speed and travels in the direction shown. Green light takes longer to get through the glass. Because of its slower speed, it is refracted as shown. Blue light travels even slower in glass. Complete the diagram by estimating the path of the blue ray.

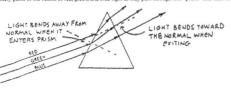

3. Below, we consider a prism-shaped hole in a piece of glass—that is, an "air prism." Complete the diagram showing likely paths of the beams of red, green, and blue light as they pass through this "prism" and back to glass.

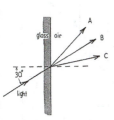

LIGHT BENDS AWAY FROM NORMAL WHEN IT ENTERS PRISM

LIGHT BENDS TOWARD THE NORMAL WHEN EXITING

85

Conceptual Integrated Science *Third Edition*

Refraction—Part 1—continued

4. Light of different colors diverges when emerging from a prism. Newton showed that with a second prism he could make the diverging beams become parallel again. Which placement of the second prism will do this?

(NOTE PARALLEL FACES!)

5. The sketch shows that due to refraction, the man sees the fish closer to the water surface than it actually is.

a. Draw a ray beginning at the fish's eye to show the line of sight of the fish when it looks upward at 50° to the normal at the water surface. Draw the direction of the ray after it meets the surface of the water.

b. At the 50° angle, does the fish see the man, or does it see the reflected view of the starfish at the bottom of the pond? Explain.

FISH SEES REFLECTED VIEW OF STARFISH (50° > 48° CRITICAL ANGLE, SO THERE IS TOTAL INTERNAL REFLECTION)

c. To see the man, should the fish look higher or lower than the 50° path?

HIGHER, SO LINE OF SIGHT TO THE WATER IS LESS THAN 48° WITH NORMAL

d. If the fish's eye were barely above the water surface, it would see the world above in a 180° view, horizon to horizon. The fish-eye view of the world above as seen beneath the water, however, is very different. Due to the 48° critical angle of water, the fish sees a normally 180° horizon-to-horizon view compressed within an angle of ___96°___.

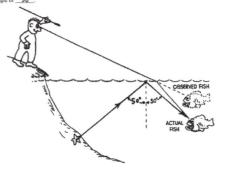

OBSERVED FISH

50° 50°

ACTUAL FISH

86

Conceptual Integrated Science *Third Edition*

Chapter 8: Waves—Sound and Light
Refraction—Part 2

1. The sketch to the right shows a light ray moving from air into water, at 45° to the normal. Which of the three rays indicated with capital letters is most likely the light ray that continues inside the water?

___C___

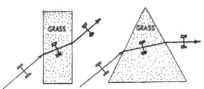

light

air
water

A

B

C

glass air

A
B
C

30°

light

2. The sketch on the left shows a light ray moving from glass into air, at 30° to the normal. Which of the three is most likely the light ray that continues in the air?

___A___

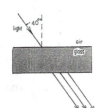

light 40°

air
glass

3. To the right, a light ray is shown moving from air into a glass block, at 40° to the normal. Which of the three rays is most likely the light ray that travels in the air after emerging from the opposite side of the block?

___A___

Sketch the path the light would take inside the glass.

light 40°

water
air

A B C

4. To the left, a light ray is shown moving from water into a rectangular block of air (inside a thin-walled plastic box), at 40° to the normal. Which of the three rays is most likely the light ray that continues into the water on the opposite side of the block?

___C___

Sketch the path the light would take inside the air.

87

Conceptual Integrated Science *Third Edition*

Refraction—Part 2—continued

5. The two transparent blocks (right) are made of different materials. The speed of light in the left block is greater than the speed of light in the right block. Draw an appropriate light path through and beyond the right block. Is the light that emerges displaced more or less than light emerging from the left block?

___MORE___

light

displacement

6. Light from the air passes through plates of glass and plastic below. The speeds of light in the different materials is shown to the right (these different speeds are often implied by the "index of refraction" of the material). Construct a rough sketch showing an appropriate path through the system of four plates.

Compared with the 50° incident ray at the top, what can you say about the angles of the ray in the air between and below the block pairs?

___SAME 50°___

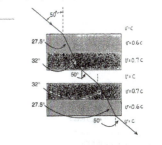

50°

$v = c$

27.5° $v = 0.6c$

32° $v = 0.7c$

50° $v = c$

32° $v = 0.7c$

27.5° $v = 0.6c$

50° $v = c$

7. Parallel rays of light are refracted as they change speed in passing from air into the eye (left). Construct a rough sketch showing appropriate light paths when parallel light under water meets the same eye (right).

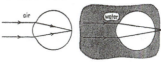

air

water

If a fish out of water wishes to clearly view objects in air, should it wear goggles filled with water or with air?

8. Why do we need to wear a face mask or goggles to see clearly when under water?

SO THAT LIGHT GOES FROM AIR TO EYE FOR PROPER REFRACTION

88

Chapter 8: Waves—Sound and Light
Wave–Particle Duality

1. To say that light is quantized means that light is made up of
(elemental units) (waves).

2. Compared with photons of low-frequency light, photons of
higher-frequency light have more
(energy) (speed) (quanta).

3. The photoelectric effect supports the
(wave model of light) (particle model of light).

4. The photoelectric effect is evident when light shone on certain
photosensitive materials ejects
(photons) (electrons).

5. The photoelectric effect is more effective with violet light than with red light
because the photons of violet light
(resonate with the atoms in the material)
(deliver more energy to the material)
(are more numerous).

6. According to the wave model of matter, a beam of light and a beam of electrons
(are fundamentally different) (are similar).

7. According to De Broglie, the greater the speed of an electron beam, the
(greater is its wavelength) (shorter is its wavelength).

8. The discreteness of the energy levels of electrons about the atomic nucleus is best understood by considering the
electron to be a
(wave) (particle).

9. Heavier atoms are not appreciably larger in size than lighter atoms. The main reason for the similarity of sizes is the
greater nuclear charge
(pulls surrounding electrons into tighter orbits)
(holds more electrons about the atomic nucleus)
(produces a denser atomic structure).

10. Whereas in the everyday macroworld the study of motion is called
mechanics, in the microworld the study of quanta is called
(Newton mechanics) (quantum mechanics).

A QUANTUM MECHANIC!

Chapter 9: Atoms and the Periodic Table
Subatomic Particles

Three fundamental particles of the atom are the __PROTON__, __NEUTRON__, and __ELECTRON__. At the center of each atom lies the atomic __NUCLEUS__, which consists of __PROTONS__ and __NEUTRONS__. The **atomic number** refers to the number of __PROTONS__ in the nucleus. All atoms of the same element have the same number of __PROTONS__, hence, the same atomic number.

Isotopes are atoms that have the same number of __PROTONS__, but a different number of __NEUTRONS__. An isotope is identified by its **atomic mass number**, which is the total number of __NEUTRONS__ and __PROTONS__ in the nucleus. A carbon isotope that has 6 __NEUTRONS__ and 6 __PROTONS__ is identified as carbon-12, where 12 is the atomic mass number. A carbon isotope having 6 __PROTONS__ and 8 __NEUTRONS__, on the other hand, is carbon-14.

1. Complete the following table:

Isotope	Number of...		
	Electrons	Protons	Neutrons
Hydrogen-1	1	1	0
Chlorine-36	17	17	19
Nitrogen-14	7	7	7
Potassium-40	19	19	21
Arsenic-75	33	33	42
Gold-197	79	79	118

2. Which results in a more valuable product—*adding* or *subtracting* protons from gold nuclei?
SUBTRACT FOR PLATINUM (MORE VALUABLE)

3. Which has more mass, a helium atom or a neon atom?
NEON

4. Which has a greater number of atoms, a gram of helium or a gram of neon?
HELIUM!

Chapter 10: The Atomic Nucleus and Radioactivity
Radioactivity

1. Complete the following statements.

a. A lone neutron spontaneously decays into a proton plus an ____ELECTRON____.

b. Alpha and beta rays are made of streams of particles, whereas gamma rays are streams of
____PHOTONS____.

c. An electrically charged atom is called an ____ION____.

d. Different ____ISOTOPES____ of an element are chemically identical but differ in the number of
neutrons in the nucleus.

e. Transuranic elements are those beyond atomic number ____92____.

f. If the amount of a certain radioactive sample decreases by half in four weeks, in four more weeks the amount
remaining should be ____¼____ the original amount.

g. Water from a natural hot spring is warmed by ____RADIOACTIVITY____ inside Earth.

2. The gas in the little girl's balloon is made up of former alpha and beta particles produced by radioactive decay.

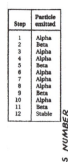

a. If the mixture is electrically neutral, how many more beta particles
than alpha particles are in the balloon?
__TWICE AS MANY BETA PARTICLES AS__
__ALPHA PARTICLES__

b. Why is your answer not "same"?
ALPHA HAS DOUBLE CHARGE; THE
CHARGE OF 2 BETAS = MAGNITUDE
OF CHARGE OF 1 ALPHA

c. Why are the alpha and beta particles no longer harmful to the child?
__THEY HAVE LOST THEIR HIGH KE, WHICH IS NOW REDUCED TO THERMAL ENERGY__
__OF RANDOM MOLECULAR MOTION.__

d. What element does this mixture make?
HELIUM

Radioactivity—continued

Draw in a decay-scheme diagram below, similar to Figure 10.17 in your text. In this case, you begin at the upper right with U-235 and end up with a different isotope of lead. Use the table at the left and identify each element in the series by its chemical symbol.

Step	Particle emitted
1	Alpha
2	Beta
3	Alpha
4	Alpha
5	Beta
6	Alpha
7	Alpha
8	Alpha
9	Beta
10	Alpha
11	Beta
12	Stable

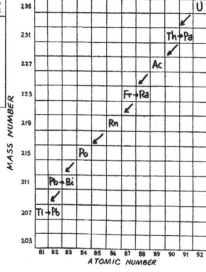

Which isotope is the final product? $^{207}_{82}$ Pb (LEAD-207)

Conceptual Integrated Science *Third Edition*

Chapter 10: The Atomic Nucleus and Radioactivity
Radioactive Half-Life

You and your classmates will now play the "half-life game." Each of you should have a coin to shake inside cupped hands. After it has been shaken for a few seconds, the coin is tossed on the table or on the floor. Students with tails up fall out of the game. Only those who consistently show heads remain in the game. Finally, everybody has tossed a tail and the game is over.

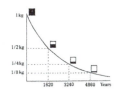

1. The graph to the left shows the decay of Radium-226 with time. Note that each 1620 years, half remains (the rest changes to other elements). In the grid below, plot the number of students left in the game after each toss. Draw a smooth curve that passes close to the points on your plot. What is the similarity of your curve with that of the curve of Radium-226?

BOTH SHOULD LOOK SIMILAR, STARTING WITH A STEEP DECENT FOLLOWED BY A LEVELING OUT

VARIES

2. Was the person to last longest in the game *lucky*, with some sort of special powers to guide the long survival? What test could you make to decide the answer to this question?

TEST! REPEAT TO SEE IF "LUCKY" PERSON REMAINS LUCKY!

Conceptual Integrated Science *Third Edition*

Chapter 10: The Atomic Nucleus and Radioactivity
Nuclear Fission and Fusion

1. Complete the table for a chain reaction in which two neutrons from each step individually cause a new reaction.

EVENT	1	2	3	4	5	6	7
NO. OF REACTIONS	1	2	4	8	16	32	

2. Complete the table for a chain reaction in which three neutrons from each reaction cause a new reaction.

EVENT	1	2	3	4	5	6	7
NO. OF REACTIONS	1	3	9	27	81	243	

3. Complete these beta reactions, which occur in a fission breeder reactor.

$$^{239}_{92}U \rightarrow ^{239}_{93}Np + ^{0}_{-1}e$$

$$^{239}_{93}Np \rightarrow ^{239}_{94}Pu + ^{0}_{-1}e$$

4. Complete the following fission reactions.

$$^{1}_{0}n + ^{235}_{92}U \rightarrow ^{143}_{54}Xe + ^{90}_{38}Sr + 3\left(^{1}_{0}n\right)$$

$$^{1}_{0}n + ^{235}_{92}U \rightarrow ^{152}_{60}Nd + ^{80}_{32}Ge + 4\left(^{1}_{0}n\right)$$

$$^{1}_{0}n + ^{239}_{94}Pu \rightarrow ^{141}_{54}Xe + ^{97}_{40}Zr + 2\left(^{1}_{0}n\right)$$

5. Complete the following fusion reactions.

$$^{2}_{1}H + ^{2}_{1}H \rightarrow ^{3}_{2}He + ^{1}_{0}n$$

$$^{2}_{1}H + ^{3}_{1}H \rightarrow ^{4}_{2}He + ^{1}_{0}n$$

Conceptual Integrated Science *Third Edition*

Chapter 10: The Atomic Nucleus and Radioactivity
Nuclear Reactions

Complete these nuclear reactions:

1. $^{230}_{90}Th \rightarrow ^{226}_{88}Ra + ^{4}_{2}He$

2. $^{218}_{85}At \rightarrow ^{214}_{83}Bi + ^{4}_{2}He$

3. $^{14}_{6}C \rightarrow ^{0}_{-1}e + ^{14}_{7}N$

4. $^{80}_{35}Br \rightarrow ^{80}_{36}Kr + ^{0}_{-1}e$

5. $^{214}_{83}Bi \rightarrow ^{4}_{2}He + ^{210}_{81}Tl$

6. $^{212}_{83}Bi \rightarrow ^{0}_{-1}e + ^{212}_{84}Po$

NUCLEAR PHYSICS··· IT'S THE SAME TO ME WITH THE FIRST TWO LETTERS INTERCHANGED!

7. $^{80}_{35}Br \rightarrow ^{0}_{-1}e + ^{80}_{36}Kr$

8. $^{80}_{35}Br \rightarrow ^{0}_{+1}e + ^{80}_{34}Se$

9. $^{1}_{1}H + ^{7}_{3}Li \rightarrow ^{4}_{2}He + ^{4}_{2}He$

10. $^{2}_{1}H + ^{3}_{1}H \rightarrow ^{4}_{2}He + ^{1}_{0}n$

Conceptual Integrated Science *Third Edition*

Chapter 11: Investigating Matter
Melting Points of the Elements

There is a remarkable degree of organization in the periodic table. As discussed in your textbook, elements within the same atomic group (vertical column) share similar properties. Also, the chemical reactivity of an element can be deduced from its position in the periodic table. Two additional examples of the periodic table's organization are the melting points and densities of the elements.

The periodic table below shows the melting points of nearly all the elements. Note the melting points are not randomly oriented, but, with only a few exceptions, either gradually increase or decrease as you move in any particular direction. This can be clearly illustrated by color coding each element according to its melting point.

Use colored pencils to color in each element according to its melting point. Use the suggested color legend. Color lightly so that symbols and numbers are still visible.

Color	Temperature Range, °C	Color	Temperature Range, °C
Violet	-273 — -50	Yellow	1400 — 1900
Blue	-50 — 300	Orange	1900 — 2900
Cyan	300 — 700	Red	2900 — 3500
Green	700 — 1400		

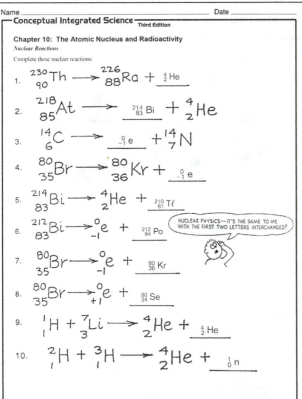

Melting Points of the Elements (°C)

TUNGSTEN

1. Which elements have the highest melting points?

THE ONES CLOSER TO TUNGSTEN

2. Which elements have the lowest melting points?

ELEMENTS TOWARD THE UPPER RIGHT

3. Which atomic groups tend to go from higher to lower melting points reading from top to bottom? (Identify each group by its group number.)

1, 2, 3, 12, 13, 14

4. Which atomic groups tend to go from lower to higher melting points reading from top to bottom?

4 THROUGH 10 AND 15 THROUGH 18

Conceptual Integrated Science
Third Edition

Chapter 11: Investigating Matter
Densities of the Elements

The periodic table below shows the densities of nearly all the elements. As with the melting points, the densities of the elements either gradually increase or decrease as you move in any particular direction. Use colored pencils to color in each element according to its density. Shown below is a suggested color legend. Color lightly so that symbols and numbers are still visible. (Note: All gaseous elements are marked with an asterisk and should be the same color. Their densities, which are given in units of g/L, are much less than the densities of nongaseous elements, which are given in units of g/mL.)

Color	Density (g/mL)	Color	Density (g/mL)
Violet	gaseous elements	Yellow	16 — 12
Blue	5 — 0	Orange	20 — 16
Cyan	8 — 5	Red.	23 — 20
Green	12 — 8		

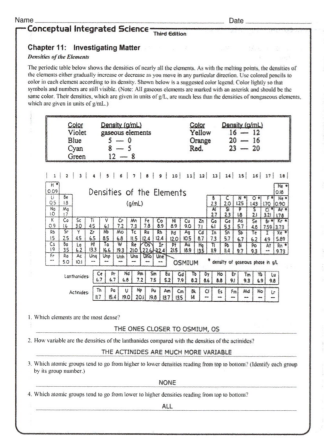

Densities of the Elements (g/mL)

OSMIUM   * density of gaseous phase in g/L

1. Which elements are the most dense?

THE ONES CLOSER TO OSMIUM, OS

2. How variable are the densities of the lanthanides compared with the densities of the actinides?

THE ACTINIDES ARE MUCH MORE VARIABLE

3. Which atomic groups tend to go from higher to lower densities reading from top to bottom? (Identify each group by its group number.)

NONE

4. Which atomic groups tend to go from lower to higher densities reading from top to bottom?

ALL

Conceptual Integrated Science
Third Edition

Chapter 11: Investigating Matter
The Submicroscopic

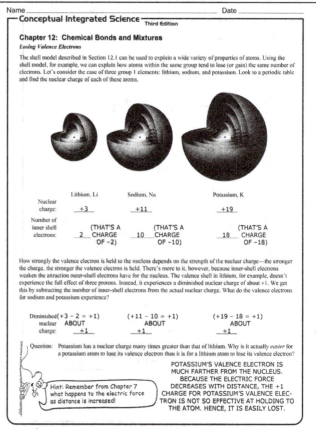

1. How many molecules are shown in   A 2   B 4   C 4
2. How many atoms are shown in   A 8   B 8   C 8
3. Which represents a physical change?   B → A   *(circle one)*
4. Which represents a chemical change?   (B → A)   B → C   *(circle one)*
5. Which box(es) represent(s) a mixture?   A ✓   B   C _____
6. Which box contains the most mass?   A ✓   B ✓   C ✓   ALL WITH SAME MASS
7. Which box is the coldest?   MAY BE WARMER OR COLDER A _____   B _____   C ✓
8. Which box contains the most air between molecules?   A NONE   B _____   C _____

THERE IS NO AIR BETWEEN THE MOLECULES.

9. How many molecules are shown in   A 2   B 3   C 2
10. How many atoms are shown in   A 6   B 6   C 6
11. Which represents a physical change?   B → A   B → C   *(circle one)*   NEITHER
12. Which represents a chemical change?   (B → A)   (B → C)   *(circle one)*   BOTH
13. Which box(es) represent(s) a mixture?   A ✓   B ✓   C _____
14. Which box contains the most mass?   A ✓   B ✓   C ✓   ALL WITH SAME MASS
15. Which should take longer?   B → A   (B → C)   *(circle one)*
ONE LESS STEP IS REQUIRED TO GO FROM B → A
16. Which box most likely contains ions?   A ✓   B _____   C _____

Conceptual Integrated Science
Third Edition

Chapter 11: Investigating Matter
Physical and Chemical Changes

Chemistry Frights

1. What distinguishes a chemical change from a physical change?
DURING A CHEMICAL CHANGE ATOMS CHANGE PARTNERS

2. On the basis of observations alone, why is distinguishing a chemical change from a physical change not always so straightforward?
BOTH INVOLVE A CHANGE IN PHYSICAL APPEARANCE

Try your hand at categorizing the following processes as either chemical or physical change. Some of these examples are debatable! Be sure to discuss your reasoning with your classmates or your instructor.

(circle one)

3. A cloud grows dark. _____ chemical (physical)
4. Leaves produce oxygen. _____ (chemical) physical
5. Food coloring is added to water. _____ chemical (physical)
6. Tropical coral reef dies. _____ (chemical) physical
7. Dead coral reef is pounded by waves into beach sand. _____ chemical (physical)
8. Oil and vinegar separate. _____ chemical (physical)
9. Soda drink goes flat. _____ chemical (physical)
10. Sick person develops a fever. _____ (chemical) physical
11. Compost pit turns into mulch. _____ (chemical) physical
12. A computer is turned on. ___AT THE ELECTRIC POWER PLANT → (chemical) (physical)
13. An electrical short melts a computer's integrated circuits. _____ chemical (physical)
14. A car battery runs down. _____ (chemical) physical
15. A pencil is sharpened. _____ chemical (physical)
16. Mascara is applied to eyelashes. _____ chemical (physical)
17. Sunbather gets tan lying in the sun. _____ (chemical) physical
18. Invisible ink turns visible upon heating. _____ (chemical) physical
19. A light bulb burns out. _____ (chemical) physical
20. Car engine consumes a tank of gasoline. _____ (chemical) physical
21. B vitamins turn urine yellow. _____ chemical (physical)
ASSUMING "XS" VITAMIN PASSES THROUGH BODY UNCHANGED

Conceptual Integrated Science
Third Edition

Chapter 12: Chemical Bonds and Mixtures
Losing Valence Electrons

The shell model described in Section 12.1 can be used to explain a wide variety of properties of atoms. Using the shell model, for example, we can explain how atoms within the same group tend to lose (or gain) the same number of electrons. Let's consider the case of three group 1 elements: lithium, sodium, and potassium. Look to a periodic table and find the nuclear charge of each of these atoms.

	Lithium, Li	Sodium, Na	Potassium, K
Nuclear charge:	+3	+11	+19
Number of inner shell electrons:	2 (THAT'S A CHARGE OF −2)	10 (THAT'S A CHARGE OF −10)	18 (THAT'S A CHARGE OF −18)

How strongly the valence electron is held to the nucleus depends on the strength of the nuclear charge—the stronger the charge, the stronger the valence electron is held. There's more to it, however, because inner-shell electrons weaken the attraction outer-shell electrons have for the nucleus. The valence shell in lithium, for example, doesn't experience the full effect of three protons. Instead, it experiences a diminished nuclear charge of about +1. We get this by subtracting the number of inner-shell electrons from the actual nuclear charge. What do the valence electrons experience for sodium and potassium?

Diminished nuclear charge:	(+3 − 2 = +1) ABOUT +1	(+11 − 10 = +1) ABOUT +1	(+19 − 18 = +1) ABOUT +1

Question: Potassium has a nuclear charge many times greater than that of lithium. Why is it actually *easier* for a potassium atom to lose its valence electron than it is for a lithium atom to lose its valence electron?

POTASSIUM'S VALENCE ELECTRON IS MUCH FARTHER FROM THE NUCLEUS. BECAUSE THE ELECTRIC FORCE DECREASES WITH DISTANCE, THE +1 CHARGE FOR POTASSIUM'S VALENCE ELECTRON IS NOT SO EFFECTIVE AT HOLDING TO THE ATOM. HENCE, IT IS EASILY LOST.

Hint: Remember from Chapter 7 what happens to the electric force as distance is increased!

Chapter 12: Chemical Bonds and Mixtures
Drawing Shells

Atomic shells can be represented by a series of concentric circles as shown in your textbook. With a little effort, however, it's possible to show these shells in three dimensions. Grab a pencil and blank sheet of paper and follow the steps shown below. Practice makes perfect.

1. Lightly draw a diagonal guideline. Then, draw a series of seven semicircles. Note how the ends of the semicircles are not perpendicular to the guideline. Instead, they are parallel to the length of the page, as shown in Figure 1.

Guideline →

Figure 1   **Figure 2**

2. Connect the ends of each semicircle with those of another semicircle such that a series of concentric hearts are drawn. The ends of these new semicircles should be drawn perpendicular to the ends of the previously drawn semicircles, as shown in Figure 2.

3. Now the hard part. Draw a portion of a circle that connects the apex of the largest vertical and horizontal semicircles, as in Figure 3.

Figure 3   **Figure 4**

4. Now the fun part. Erase the pencil guideline then add the internal lines, as shown in Figure 4, that create a series of concentric shells.

You need not draw all the shells for each atom. Oxygen, for example, is nicely represented drawing only the first two inner shells, which are the only ones that contain electrons. Remember that these shells are not to be taken literally. Rather, they are a highly simplified view of how electrons tend to organize themselves within an atom. You should know that each shell represents a set of atomic orbitals of similar energy levels as shown in your textbook.

Chapter 12: Chemical Bonds and Mixtures
Atomic Size

1. Complete the shells for the following atoms using arrows to represent electrons.

Li   Be   B   C   N   O   F   Ne

2. Neon, Ne, has many more electrons than lithium, Li, yet it is a much smaller atom. Why?

NEON HAS A STRONGER NUCLEAR CHARGE (+10) THAT PULLS THE ELECTRONS IN CLOSER TO IT.

3. Draw the shell model for a sodium atom, Na (atomic number 11), adjacent to the neon atom in the box shown below. Use a pencil because you may need to erase.

Ne   Na

a. Which should be larger, neon's first shell or sodium's first shell? Why? Did you represent this accurately within your drawing?

NEON'S FIRST SHELL IS LARGER BECAUSE OF THE WEAKER NUCLEAR CHARGE.

b. Which has a greater nuclear charge, Ne or Na?

SODIUM, Na

c. Which is a larger atom, Ne or Na?

SODIUM, BUT NOT BECAUSE OF A GREATER NUCLEAR CHARGE, BUT BECAUSE OF THE EXTRA SHELL OF ELECTRONS.

4. Moving from left to right across the periodic table, what happens to the nuclear charge within atoms? What happens to atomic size?

THE NUCLEAR CHARGE INCREASES FROM LEFT TO RIGHT ACROSS THE PERIODIC TABLE, WHICH IS WHY THE ATOMIC SIZE DECREASES.

5. Moving from top to bottom down the periodic table, what happens to the number of occupied shells? What happens to atomic size?

MOVING DOWN A GROUP, THE NUMBER OF OCCUPIED SHELLS INCREASES, WHICH IS WHY THE ATOMIC SIZE ALSO INCREASES.

6. Where in the periodic table are the smallest atoms found? Where are the largest atoms found?

THE SMALLEST ATOMS ARE FOUND TO THE UPPER RIGHT WHILE THE LARGEST ATOMS ARE FOUND TO THE LOWER LEFT.

Chapter 12: Chemical Bonds and Mixtures
Effective Nuclear Charge

 The magnitude of the nuclear charge sensed by an orbiting electron depends upon several factors, including the number of positively charged protons in the nucleus, the number of inner shell electrons shielding it from the nucleus, and its distance from the nucleus.

1. Place the proper number of electrons in each shell for carbon and silicon (use arrows to represent electrons).

IT'S CLOSER   Carbon   Silicon

2. According to the shell model, which should experience the greater effective nuclear charge, an electron in

a. carbon's 1st shell or silicon's 1st shell?   (circle one)

b. carbon's 2nd shell or silicon's 2nd shell?   (circle one)

c. carbon's 2nd shell or silicon's 3rd shell?   (circle one)

3. List the shells of carbon and silicon in order of decreasing effective nuclear charge.

SILICON'S 1ST	>	SILICON'S 2ND	>	CARBON'S 1ST	>	CARBON'S 2ND	>	SILICON'S 3RD
~+14		~+12		~+6		~+4		~≤+4

4. Which should have the greater ionization energy, the carbon atom or the silicon atom? Defend your answer.   THE CARBON ATOM, BECAUSE ITS OUTERMOST ELECTRON IS EXPERIENCING A GREATER EFFECTIVE NUCLEAR CHARGE.

5. How many additional electrons are able to fit in the outermost shell of carbon? 4 Silicon? 4

6. Which should be stronger, a C-H bond or an Si-H bond? Defend your answer.
C-H IS STRONGER. THEIR VALENCE ELECTRONS EXPERIENCE A GREATER EFFECTIVE NUCLEAR CHARGE, HENCE, THEY ARE HELD TIGHTER.

7. Which should be larger in size, the ion C^{4+} or the ion Si^{4+}? Why?
THE ARRANGEMENTS OF ELECTRONS ARE THE SAME.
SILICON, HOWEVER, HAS A GREATER NUCLEAR CHARGE, WHICH PULLS ELECTRONS INWARD MAKING Si^{4+} SMALLER.

Chapter 12: Chemical Bonds and Mixtures
Solutions

1. Use these terms to complete the following sentences. Some terms may be used more than once.

solution	solvent	solute
dissolve	concentrated	dilute
saturated	concentration	mole
molarity	solubility	soluble
insoluble	precipitate	

Sugar is SOLUBLE in water for the two can be mixed homogeneously to form a SOLUTION . The SOLUBILITY of sugar in water is so great that CONCENTRATED homogeneous mixtures are easily prepared. Sugar, however, is not infinitely SOLUBLE in water, for when too much of this SOLUTE is added to water, which behaves as the SOLVENT , the solution becomes SATURATED . At this point any additional sugar is INSOLUBLE for it will not DISSOLVE . If the temperature of a saturated sugar solution is lowered, the SOLUBILITY of the sugar in water is also lowered. If some of the sugar comes out of solution, it is said to form a PRECIPITATE . If, however, the sugar remains in solution despite the decrease in solubility, then the solution is said to be supersaturated. Adding only a small amount of sugar to water results in a DILUTE solution. The CONCENTRATION of this solution or any solution can be measured in terms of MOLARITY , which tells us the number of solute molecules per liter of solution. If there are 6.022×10^{23} molecules in 1 liter of solution, then the CONCENTRATION of the solution is 1 MOLE per liter.

2. Temperature has a variety of effects on the solubilities of various solutes. With some solutes, such as sugar, solubility increases with increasing temperature. With other solutes, such as sodium chloride (table salt), changing temperature has no significant effect. With some solutes, such as lithium sulfate (Li_2SO_4) the solubility actually decreases with increasing temperature.

a. Describe how you would prepare a supersaturated solution of lithium sulfate.
FORM A SATURATED SOLUTION AND THEN SLOWLY RAISE THE TEMPERATURE

b. How might you cause a saturated solution of lithium sulfate to form a precipitate?
INCREASE ITS TEMERATURE

Conceptual Integrated Science — Third Edition

Chapter 12: Chemical Bonds and Mixtures
Pure Mathematics

Using a scientist's definition of *pure*, identify whether each of the following is 100% pure:

	100% pure?	
Freshly squeezed orange juice	Yes	(No)
Country air	Yes	(No)
Ocean water	Yes	(No)
Fresh drinking water	Yes	(No)
Skim milk	Yes	(No)
Stainless steel	Yes	(No)
A single water molecule	(Yes)	No

A glass of water contains in the order of a trillion trillion (1×10^{24}) molecules. If the water in this were 99.9999% pure, you could calculate the percentage of impurities by subtracting from 100.0000%.

$$100.0000\% \text{ water + impurity molecules}$$
$$- \ 99.9999\% \text{ water molecules}$$
$$\overline{\quad 0.0001\% \text{ impurity molecules}}$$

Pull out your calculator and calculate the number of impurity molecules in the glass of water. Do this by finding 0.0001% of 1×10^{24}, which is the same as multiplying 1×10^{24} by 0.000001.

$$(1 \times 10^{24})(0.000001) = \underline{\quad 1 \times 10^{18} \quad}$$

1. How many impurity molecules are there in a glass of water that's 99.9999% pure?

   a. 1000 (one thousand: 10^3)

   b. 1,000,000 (one million: 10^6)

   c. 1,000,000,000 (one billion: 10^9)

   ⓓ. 1,000,000,000,000,000,000 (one million trillion: 10^{18})

2. How does your answer make you feel about drinking water that is 99.9999 % free of some poison such as pesticide?
   THAT THERE ARE A MILLION TRILLION POISON MOLECULES IN A GLASS OF WATER MIGHT MAKE ONE HESITATE . . . BUT READ ON!

3. For every one impurity molecule, how many water molecules are there? (Divide the number of water molecules by the number of impurity molecules.)
   $$10^{24}/10^{18} = 10^6 = 1,000,000 = \text{one million}$$

4. Would you describe these impurity molecules within water that's 99.9999% pure as "rare" or "common"?
   FOR EVERY ONE IMPURITY MOLECULE THERE ARE ONE MILLION WATER MOLECULES. ONE IN A MILLION IS RARE!

5. A friend argues that he or she doesn't drink tap water because it contains thousands of molecules of some impurity in each glass. How would you respond in defense of the water's purity, if it indeed does contain thousands of molecules of some impurity per glass?

   ONLY 1,000 IMPURITY MOLECULES IN THIS GLASS OF WATER WOULD MAKE THIS WATER INCREDIBLY PURE . . . ABOUT 99.999999999999999999% PURE!

Conceptual Integrated Science — Third Edition

Chapter 12: Chemical Bonds and Mixtures
Chemical Bonds

1. On the basis of their positions in the periodic table, predict whether each pair of elements will form an ionic bond, covalent bond, or neither (atomic number in parenthesis).

   a. Gold (79) and platinum (78) __N__   f. Germanium (32) and arsenic (33) __I__

   b. Rubidium (37) and iodine (53) __C__   g. Iron (26) and chromium (24) __I__

   c. Sulfur (16) and chlorine (17) __I__   h. Chlorine (17) and iodine (53) __C__

   d. Sulfur (16) and magnesium (12) __N__   i. Carbon (6) and bromine (35) __C__

   e. Calcium (20) and chlorine (17) __C__   j. Barium (56) and astatine (85) __I__

2. The most common ions of lithium, magnesium, aluminum, chlorine, oxygen, and nitrogen and their respective charges are as follows:

Positively Charged Ions	Negatively Charged Ions
Lithium ion: Li^{1+}	Chloride ion: Cl^{1-}
Barium ion: Ba^{2+}	Oxide ion: O^{2-}
Aluminum ion: Al^{3+}	Nitride ion: N^{3-}

   Use this information to predict the chemical formulas for the following ionic compounds:

   a. Lithium chloride: $LiCl$ d. Lithium oxide: Li_2O g. Lithium nitride: Li_3N

   b. Barium chloride: $BaCl_2$ e. Barium oxide: BaO h. Barium nitride: Ba_3N_2

   c. Aluminum chloride: $AlCl_3$ f. Aluminum oxide: Al_2O_3 i. Aluminum nitride: AlN

   j. How are elements that form positive ions grouped in the periodic table relative to elements that form negative ions? __POSITIVE ION ELEMENTS TOWARD THE LEFT AND NEGATIVE IONS TOWARD THE RIGHT__

3. Specify whether the following chemical structures are polar or nonpolar:

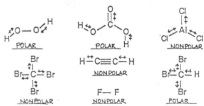

POLAR   POLAR   NONPOLAR

NONPOLAR   NONPOLAR   POLAR

Conceptual Integrated Science — Third Edition

Chapter 12: Chemical Bonds and Mixtures
Shells and the Covalent Bond

When atoms bond covalently, their atomic shells overlap so that shared electrons can occupy both shells at the same time.

Nonbonded hydrogen atoms   Covalently bonded hydrogen atoms

Hydrogen   Hydrogen   Molecular Hydrogen
Formula: H_2

Fill each shell model shown below with enough electrons to make each atom electrically neutral. Use arrows to represent electrons. Within the box draw a sketch showing how the two atoms bond covalently. Draw hydrogen shells more than once when necessary so that no electrons remain unpaired. Write the name and chemical formula for each compound.

A.

Hydrogen   Carbon

Name of Compound: METHANE Formula: CH_4

B.

Hydrogen   Nitrogen

Name of Compound: AMMONIA Formula: NH_3

Conceptual Integrated Science — Third Edition

Shells and the Covalent Bond—continued

C.

Hydrogen   Oxygen

Name of Compound: WATER Formula: H_2O

D.

Hydrogen   Fluorine

Name of Compound: HYDROGEN FLUORIDE Formula: HF

E.

Hydrogen   Neon

BONDING NOT POSSIBLE!

Name of Compound: _____ Formula: _____

1. Note the relative positions of carbon, nitrogen, oxygen, fluorine, and neon in the periodic table. How does this relate to the number of times each of these elements is able to bond with hydrogen?

   IT'S IN A DESCENDING ORDER FROM LEFT TO RIGHT.

2. How many times is the element boron (atomic number 5) able to bond with hydrogen? Use the shell model to help you with your answer.

   ONLY 3 VALENCE ELECTRONS, THEREFORE, ONLY 3 BONDS

300

Chapter 12: Chemical Bonds and Mixtures
Bond Polarity

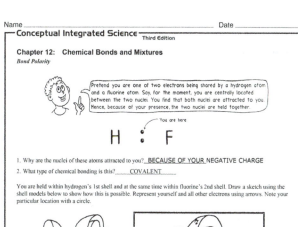

Pretend you are one of two electrons being shared by a hydrogen atom and a fluorine atom. Say, for the moment, you are centrally located between the two nuclei. You find that both nuclei are attracted to you. Hence, because of your presence, the two nuclei are held together.

You are here

H : F

1. Why are the nuclei of these atoms attracted to you? __BECAUSE OF YOUR NEGATIVE CHARGE__

2. What type of chemical bonding is this? ____COVALENT____

You are held within hydrogen's 1st shell and at the same time within fluorine's 2nd shell. Draw a sketch using the shell models below to show how this is possible. Represent yourself and all other electrons using arrows. Note your particular location with a circle.

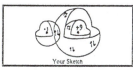

Hydrogen   Fluorine

Your Sketch

According to the laws of physics, if the nuclei are both attracted to you, then you are attracted to both of the nuclei.

3. You are pulled toward the hydrogen nucleus, which has a positive charge. How strong is this charge from your point of view—what is its *electronegativity*? __~+1__

4. You are also attracted to the fluorine nucleus. What is its electronegativity? __~+7__

You are being shared by the hydrogen and fluorine nuclei. But as a moving electron you have some choice as to your location.

5. Consider the electronegativities you experience from both nuclei. Which nucleus would you tend to be closest to? __FLUORINE__

125

Bond Polarity—continued

Stop pretending you are an electron and observe the hydrogen-fluorine bond from outside the hydrogen fluoride molecule. Bonding electrons tend to congregate to one side because of the differences in effective nuclear charges. This makes one side slightly negative in character and the opposite side slightly positive. Indicate this on the following structure for hydrogen fluoride using the symbols δ- and δ+

H : F

By convention, bonding electrons are not shown. Instead, a line is simply drawn connecting the two bonded atoms. Indicate the slightly negative and positive ends.

H — F

6. Would you describe hydrogen fluoride as a polar or nonpolar molecule? __POLAR__

7. If two hydrogen fluoride molecules were thrown together, would they stick or repel? (Hint: What happens when you throw two small magnets together?) __STICK__

8. Place bonds between the hydrogen and fluorine atoms to show many hydrogen fluoride molecules grouped together. Each element should be bonded only once. Circle each molecule and indicate the slightly negative and slightly positive ends.

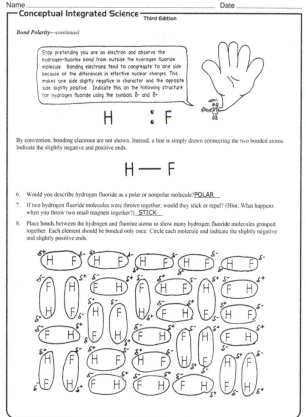

126

Chapter 12: Chemical Bonds and Mixtures
Atoms to Molecules

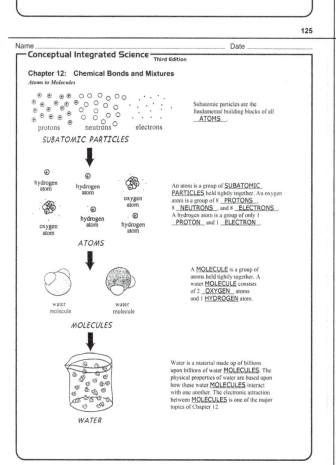

Subatomic particles are the fundamental building blocks of all __ATOMS__.

An atom is a group of __SUBATOMIC PARTICLES__ held tightly together. An oxygen atom is a group of 8 __PROTONS__, 8 __NEUTRONS__, and 8 __ELECTRONS__. A hydrogen atom is a group of only 1 __PROTON__ and 1 __ELECTRON__.

A __MOLECULE__ is a group of atoms held tightly together. A water __MOLECULE__ consists of 2 __OXYGEN__ atoms and 1 __HYDROGEN__ atom.

Water is a material made up of billions upon billions of water __MOLECULES__. The physical properties of water are based upon how these water __MOLECULES__ interact with one another. The electronic attraction between __MOLECULES__ is one of the major topics of Chapter 12.

127

Chapter 12: Chemical Bonds and Mixtures
Electron-Dot Structures

Electron-dot structures tell us how atoms tend to bond with other atoms. Carbon's electron-dot structure, for example, shows 4 unpaired valence electrons. Each hydrogen atom has 1 unpaired electron. Unpaired electrons from different atoms can pair up, resulting in a bond. For carbon and hydrogen, we have the following, which creates the molecule methane, CH4:

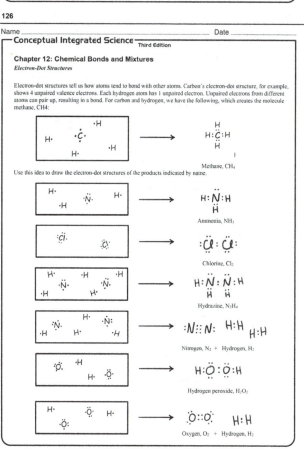

Methane, CH_4

Use this idea to draw the electron-dot structures of the products indicated by name.

Ammonia, NH_3

Chlorine, Cl_2

Hydrazine, N_2H_4

Nitrogen, N_2 + Hydrogen, H_2

Hydrogen peroxide, H_2O_2

Oxygen, O_2 + Hydrogen, H_2

129

301

┌─ **Conceptual Integrated Science** ─ Third Edition

Chapter 12: Chemical Bonds and Mixtures
Molarity

Concentration is a measure of the amount of solute within a solution. For the solute, chemists often go by a count of the number of solute particles. For the solution, they usually consider the volume of the solution in units of liters. Note, the volume of solution is the combined volume of the solute and the solvent.

For example, a chemist drops 5 marbles into a beaker of water to make a total of 0.5 liters. The concentration would be:

$$\frac{5\ moles}{0.5\ \mathcal{L}} = 10 \quad \text{marbles per liter}$$
↳ plug in answer

Of course, marbles don't dissolve in water, but they do serve to illustrate the point that it's not just the solvent that contributes to the volume of a solution. Molecules are much, much smaller than marbles so we count them not by the single, not by the dozen, but by the mole, which is an astronomically large number: 6.02×10^{23}. This many sugar molecules happen to have a mass of 342 grams.

A chemist drops 684 grams of sugar into a beaker of water to make a total of 1.5 liters of solution. After the sugar dissolves, what is the concentration?

amount of sugar (in moles) ↘
volume of solution (in $\mathcal{L}$) ↗

$$\frac{(2\ moles)}{(1.5\ \mathcal{L})} = 1.33 \quad \text{moles per liter}$$
↳ plug in answer

In the language of chemistry, "moles per liter" is also called "molar." So when a chemist wants to know the concentration of a solution, she'll ask, "What's the molarity?"

What is the molarity of the following solutions?

$$\frac{(171\ sugar)}{(3\mathcal{L}\ soln)} = \frac{(0.5\ mole)}{(3\ \mathcal{L})} = 0.17 \ molar \qquad \frac{(342\ sugar)}{(1\ \mathcal{L}\ soln)} = \frac{(1\ mole)}{(1\ \mathcal{L})} = 1 \ molar$$

Molar is usually abbreviated with a capital M. So a 4.2 molar solution of sugar water may be expressed as 4.2 M.

1. So how many grams of sugar are there in a 3 M solution of sugar water? Please explain.

   Not enough information is given to provide an answer. What's missing is the volume of solution. Clearly, if you had a swimming pool full of a 3 M solution, you would have more sugar than if you had a cup full of a 3 M solution.

2. There are two things you need to know in order to calculate the amount of solute in a solution. What are these two things?

   You need the concentration of the solution as well as the volume of solution.

┌─ **Conceptual Integrated Science** ─ Third Edition

Chapter 12: Chemical Bonds and Mixtures
Polyatomic Ions

Sometimes a molecule can lose or gain a proton (hydrogen ion) to form what we call a polyatomic ion:

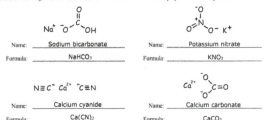

Phosphoric acid (molecule) → Phosphate ion (polyatomic ion) + H^+
Ammonia (molecule) + H^+ → Ammonium ion (polyatomic ion)

Table of common polyatomic ions

NAME	FORMULA	NAME	FORMULA
Ammonium ion	NH_4^+	Hydroxide ion	OH^-
Bicarbonate ion	HCO_3^-	Nitrate ion	NO_3^-
Carbonate ion	CO_3^{2-}	Phosphate ion	PO_4^{3-}
Cyanide ion	CN^-	Sulfate ion	SO_4^{2-}

When it comes to naming compounds, a polyatomic ion is treated as a single unit. Positively charged ions are listed first followed by the negatively charged ions, but we don't include the word "ion." For example, below is the formula for ammonium phosphate. Note how we need three (1+) ammoniums to balance a single (3-) phosphate.

positively charged ion ↘   ↙ negatively charged ion
$(NH_4)_3\ PO_4$

Use the table of common polyatomic ions to deduce the formula for the following compounds:

Ammonium sulfate ___(NH₄) SO₄___   Potassium cyanide ___KCN___

Sodium sulfate ___Na₂ SO₄___   Calcium phosphate ___Ca₃(PO₄)₂___

Sodium hydroxide ___Na OH___   Aluminum hydroxide ___Al(OH)₃___

Hydrogen hydroxide ___HOH___   Aluminum sulfate ___Al₂(SO₄)₃___
                     water!

Name the following structures and write their formula on the basis of the polyatomic ions they contain:

Name: ___Sodium bicarbonate___   Name: ___Potassium nitrate___
Formula: ___NaHCO₃___   Formula: ___KNO₃___

$N\equiv C^-\ Ca^{2+}\ ^-C\equiv N$ Ca^{2+} carbonate

Name: ___Calcium cyanide___   Name: ___Calcium carbonate___
Formula: ___Ca(CN)₂___   Formula: ___CaCO₃___

┌─ **Conceptual Integrated Science** ─ Third Edition

Chapter 13: Chemical Reactions
Balancing Chemical Equations

In a balanced chemical equation the number of times each element appears as a reactant is equal to the number of times it appears as a product. For example,

$$2\ H_2 + O_2 \longrightarrow 2\ H_2O$$

Recall that *coefficients* (the integer appearing before the chemical formula) indicate the number of times each chemical formula is to be counted and *subscripts* indicate when a particular element occurs more than once within the formula.

Check whether the following chemical equations are balanced:

$3\ NO \longrightarrow N_2O + NO_2$ ☑ balanced ☐ unbalanced

$SiO_2 + 4\ HF \longrightarrow SiF_4 + 2\ H_2O$ ☑ balanced ☐ unbalanced

$4\ NH_3 + 5\ O_2 \longrightarrow 4\ NO + 6\ H_2O$ ☑ balanced ☐ unbalanced

Unbalanced equations are balanced by changing the coefficients. Subscripts, however, should never be changed because this changes the chemical's identity—H_2O is water, but H_2O_2 is hydrogen peroxide! The following steps may help guide you:

1. Focus on balancing only one element at a time. Start with the left-most element and modify the coefficients such that this element appears on both sides of the arrow the same number of times.

2. Move to the next element and modify the coefficients so as to balance this element. Do not worry if you incidentally unbalance the previous element. You will come back to it in subsequent steps.

3. Continue from left to right, balancing each element individually.

4. Repeat steps 1–3 until all elements are balanced.

Use the above methodology to balance the following chemical equations:

$2\ N_2O \longrightarrow 2\ N_2 + 1\ O_2$

$2\ NaClO_3 \longrightarrow 2\ NaCl + 3\ O_2$

$3\ MnCl_2 + 2\ Al \longrightarrow 3\ Mn + 2\ AlCl_3$

$2\ K + 2\ H_2O \longrightarrow 1\ H_2 + 2\ KOH$

$2\ Al_2O_3 + 3\ C \longrightarrow 4\ Al + 3\ CO_2$

$4\ NH_3 + 3\ F_2 \longrightarrow 3\ NH_4F + 1\ NF_3$

This is just one of the many methods that chemists have developed to balance chemical equations.

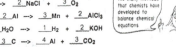

Knowing how to balance a chemical equation is a useful technique, but understanding *why* a chemical equation needs to be balanced in the first place is far more important.

┌─ **Conceptual Integrated Science** ─ Third Edition

Chapter 13: Chemical Reactions
Exothermic and Endothermic Reactions

During a chemical reaction atoms are neither created nor destroyed. Instead, atoms rearrange—they change partners. This rearrangement of atoms necessarily involves the input and output of energy. First, energy must be supplied to break chemical bonds that hold atoms together. Separated atoms then form new chemical bonds, which involves the release of energy. In an **exothermic** reaction more energy is released than is consumed. Conversely, in an **endothermic** reaction more energy is consumed than is released.

Table 1 Bond Energies

Bond	Bond Energy*	Bond	Bond Energy*
H—H	436	Cl—Cl	243
H—C	414	N—N	159
H—N	389	O=O	498
H—O	464	O=C	803
H—Cl	431	N≡N	946

*In kJ/mol

Table 1 shows bond energies—the amount of energy required to break a chemical bond, and also the amount of energy released when a bond is formed. Use these bond energies to determine whether the following chemical reactions are exothermic or endothermic:

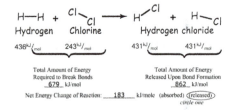

$H—H$ + $Cl—Cl$ → $H—Cl$ + $H—Cl$
Hydrogen Chlorine   Hydrogen chloride

$436^{kJ}/_{mol}$ $243^{kJ}/_{mol}$ $431^{kJ}/_{mol}$ $431^{kJ}/_{mol}$

Total Amount of Energy
Required to Break Bonds
___679___ kJ/mol

Total Amount of Energy
Released Upon Bond Formation
___862___ kJ/mol

Net Energy Change of Reaction: ___183___ kJ/mole (absorbed / released)
circle one

1. Is this reaction exothermic or endothermic? EXOTHERMIC

2. Write the balanced equation for this reaction using chemical formulas and coefficients. If it is exothermic, write "Energy" as a product. If it is endothermic, write "Energy" as a reactant.

$$H_2 + Cl_2 \rightarrow 2\ HCl + \text{ENERGY}$$

Exothermic and Endothermic Reactions—continued

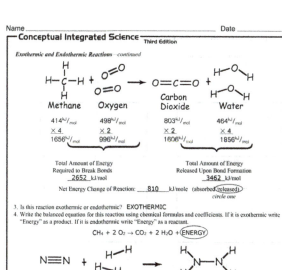

Methane	Oxygen		Carbon Dioxide	Water
$414^{kJ}/_{mol}$	$498^{kJ}/_{mol}$		$803^{kJ}/_{mol}$	$464^{kJ}/_{mol}$
$\times\,4$	$\times\,2$		$\times\,2$	$\times\,4$
$1656^{kJ}/_{mol}$	$996^{kJ}/_{mol}$		$1606^{kJ}/_{mol}$	$1856^{kJ}/_{mol}$

Total Amount of Energy Required to Break Bonds ___2652___ kJ/mol

Total Amount of Energy Released Upon Bond Formation ___3462___ kJ/mol

Net Energy Change of Reaction: ___810___ kJ/mole (absorbed/(released))
circle one

3. Is this reaction exothermic or endothermic? **EXOTHERMIC**

4. Write the balanced equation for this reaction using chemical formulas and coefficients. If it is exothermic write "Energy" as a product. If it is endothermic write "Energy" as a reactant.

$$CH_4 + 2\,O_2 \rightarrow CO_2 + 2\,H_2O + \boxed{ENERGY}$$

Nitrogen	Hydrogen		Hydrazine	
$946^{kJ}/_{mol}$	$436^{kJ}/_{mol}$		$389^{kJ}/_{mol}$	$159^{kJ}/_{mol}$
	$\times\,2$		$\times\,4$	
	$872^{kJ}/_{mol}$		$1556^{kJ}/_{mol}$	

Total Amount of Energy Required to Break Bonds ___1818___ kJ/mol

Total Amount of Energy Released Upon Bond Formation ___1715___ kJ/mol

Net Energy Change of Reaction: ___103___ kJ/mole ((absorbed)/released)
circle one

5. Is this reaction exothermic or endothermic? **EXOTHERMIC**

6. Write the balanced equation for this reaction using chemical formulas and coefficients. If it is exothermic, write "Energy" as a product. If it is endothermic, write "Energy" as a reactant.

$$\boxed{ENERGY} + N_2 + 2\,H_2 \rightarrow N_2H_4$$

Chapter 13: Chemical Reactions
Donating and Accepting Hydrogen Ions

A chemical reaction that involves the transfer of a hydrogen ion from one molecule to another is classified as an acid–base reaction. The molecule that donates the hydrogen ion behaves as an acid. The molecule that accepts the hydrogen ion behaves as a base.

On paper, the acid–base process can be depicted through a series of frames:

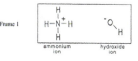

Frame 1 — Ammonium and hydroxide ions in close proximity

Frame 2 — Bond is broken between the nitrogen and a hydrogen of the ammonium ion. The two electrons of the broken bond stay with the nitrogen leaving the hydrogen with a positive charge.

Frame 3 — The hydrogen ion migrates to the hydroxide ion.

Frame 4 — The hydrogen ion bonds with the hydroxide ion to form a water molecule.

In equation form we abbreviate this process by only showing the before and after:

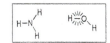

frame 1   frame 4

Donating and Accepting Hydrogen Ions—continued

We see from the previous reaction that because the ammonium ion donated a hydrogen ion, it behaved as an acid. Conversely, the hydroxide ion by accepting a hydrogen ion behaved as a base. How do the ammonia and water molecules behave during the reverse process?

acid	base		BASE	ACID
			ammonia	water

Identify the following molecules as behaving as an acid or a base:

ACID	BASE		BASE	ACID

ACID	BASE		BASE	ACID

H–H	⁻H		H⁻	H–H
ACID	BASE		BASE	ACID

HNO_3	NH_3		$^-NO_3$	$^+NH_4$
ACID	BASE		BASE	ACID

Chapter 13: Chemical Reactions
Loss and Gain of Electrons

A chemical reaction that involves the transfer of an electron is classified as an oxidation–reduction reaction. Oxidation is the process of losing electrons, while reduction is the process of gaining them. Any chemical that causes another chemical to lose electrons (become oxidized) is called an *oxidizing agent*. Conversely, any chemical that causes another chemical to gain electrons is called a *reducing agent*.

1. What is the relationship between an atom's ability to behave as an oxidizing agent and its electron affinity?
   THE GREATER THE ELECTRON AFFINITY, THE GREATER ITS ABILITY TO BEHAVE AS AN OXIDIZING AGENT.

2. Relative to the periodic table, which elements tend to behave as strong oxidizing agents?
   THOSE TO THE UPPER RIGHT WITH THE EXCEPTION OF THE NOBLE GASES.

3. Why don't the noble gases behave as oxidizing agents?
   THEY HAVE NO SPACE IN THEIR SHELLS TO ACCOMODATE ADDITIONAL ELECTRONS.

4. How is it that an oxidizing agent is itself reduced?
   REDUCTION IS THE GAINING OF ELECTRONS. IN PULLING AN ELECTRON AWAY FROM ANOTHER ATOM AN OXIDIZING AGENT NECESSARILY GAINS AN ELECTRON.

5. Specify whether each reactant is about to be oxidized or reduced.

   $$2\,K + H_2O \rightarrow 2\,K^+ + {}^-OH$$
   OX   RED

   $$2\,Mg + O_2 \rightarrow 2\,Mg^{2+}O^{2-}$$
   OX   RED

   $$2\,Na + Cl_2 \rightarrow 2\,Na^+Cl^-$$
   OX   RED

   $$CH_4 + 2\,O_2 \rightarrow O{=}C{=}O + H\!-\!O\!-\!H$$
   OX   RED

6. Which oxygen atom enjoys a greater negative charge?

   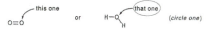

   this one   or   (that one)   *(circle one)*

7. Relate your answer to Question 6 to how it is that O_2 is reduced upon reacting with CH_4 to form carbon dioxide and water.
   IN TRANSFORMING FROM O_2 TO H_2O, AN OXYGEN ATOM IS _GAINING_ ELECTRONS AS BEST AS IT CAN. WITH ITS GREATER NEGATIVE CHARGE IT CAN BE THOUGHT OF AS "REDUCED."

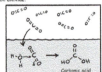

Conceptual Integrated Science *Third Edition*

Chapter 13: Chemical Reactions
Ocean Acidification

Because carbon dioxide is a non polar gas, you might expect that it does not mix readily with water, which is a polar liquid. However, as soon as carbon dioxide enters water, it reacts to form a new substance called carbonic acid, which has much better water solubility. Because of this chemistry, water is able to absorb relatively large quantities of carbon dioxide.

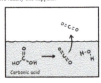

1. What happens to the pH of water as more CO_2 molecules are drawn into it?

   More CO_2 draw into water means greater amounts of carbonic acid, which causes the pH of the water to decrease (become more acidic).

Carbonic acid, in turn, can transform back into carbon dioxide as shown below. In general, the higher the temperature, the more readily this happens.

2. Why does seltzer water have such a low pH? (around 3.5!)

   This is a function of the solubility of carbonic acid and its ability to react with water to form very water-soluble ions. If this solubility and reactivity were greater, then the pH could be driven even lower.

3. Why does warm seltzer water go flat faster than cold seltzer water?

   At higher temperatures, carbon dioxide is driven out of solution at a faster rate. Relative to our oceans, higher global temperatures mean the oceans are less able to retain the carbon dioxide they absorb.

There's even more chemistry when it comes to our oceans, which are alkaline because of dissolved compounds such as calcium carbonate, $CaCO_3$.

4. Does the alkalinity of the oceans help or hinder the absorption of carbon dioxide? Please explain.

   With the alkalinity, the carbonic acid reacts to form water-soluble salts that can precipitate and sink to the ocean floor. The alkalinity definitely helps the ocean's ability to absorb CO_2.

5. What happens to the pH of the oceans as they absorb increasing amounts of carbon dioxide?

   The pH of the oceans will fall. Over the past century the oceans' pH has decreased from 8.2 to 8.1. It is expected to drop below 8.0 sometime this current century as atmospheric CO_2 continues to climb.

6. How is a coral reef connected to a coal-fired power plant?

   The predominate source of human-made carbon dioxide affecting the pH of the ocean is coal-fired power plants. The chemical effects on the ocean are quite predictable. The effects on the ocean's ecology, much less so.

7. If the oceans can absorb carbon dioxide, why have the atmospheric levels of carbon dioxide been rising so rapidly over the past century?

   About 30% of the CO_2 we place into the atmosphere makes its way to the ocean. The ocean's ability to absorb CO_2 is considerable, but it's also limited. Apparently, the rate at which we produce CO_2 is greater than the ocean's ability to absorb it. Ocean acidification is said to be "climate change's equally destructive twin."

141

Chapter 13: Chemical Reactions
Fuel Cells

The reaction of 4 hydroxide ions, HO^-, with 2 hydrogen molecules, H_2, creates 4 water molecules, H_2O, plus 4 electrons, e^-:

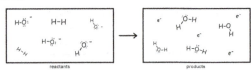

In a fuel cell this can be coupled with a second reaction in which 4 electrons, e^-, react with an oxygen molecule, O_2, and 2 water molecules, H_2O, to form 4 hydroxide ions, HO^-:

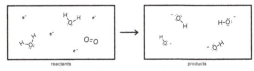

Combine all the reactants from the above two equations into the box shown below on the left and all the products from above into the box on the right to create a single big equation:

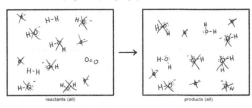

Each of these reactants and products appears at some time during the operation of this type of hydrogen–oxygen fuel cell. But what is the net result? To find out, cross out each reactant on the left also appearing as a product on the right. For example, the two reactant water molecules should be crossed out along with two of the product water molecules.

1. Write the net equation you come up with after crossing out duplicate chemicals.

$$H-H + H-H + O=O \longrightarrow H-O-H \quad H-O-H \quad +Energy$$

142

---

2. Is this reaction endothermic or exothermic? How does the light bulb become lit?

   As per the chapter section on endo- and exothermic reactions, the formation of water from oxygen and hydrogen is exothermic (Energy releasing). Interestingly, it's the energy of the released electrons that lights up the light bulb, not the electrons themselves.

3. Use the following graphic to explain how the hydroxide ions you crossed out in the products eventually become hydroxide ions of the reactants.

   Hydroxide ions formed at the cathode physically migrate across the paste to reach the anode, where they serve as reactants for the formation of water. So, although there is no net formation of hydroxide ions, they nonetheless are formed and play the important role of completing the electric circuit.

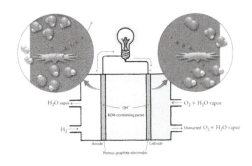

4. The anode provides electrons to the cathode. Will these electrons keep flowing if the oxygen supply is cut off? Please explain.

   Recall that oxygen atoms have a strong affinity for electrons. It's this affinity that, figuratively speaking, "pulls" the electrons toward the cathode. If the supply of oxygen were stopped, so would the flow of electrons.

5. The cathode collects electrons from the anode. Will this continue if the hydrogen supply is cut off? Please explain.

   The hydrogen serves as the source of electrons. If the supply of hydrogen were stopped, then there would be no more electrons to flow into and through the anode.

143

Chapter 13: Chemical Reactions
Iodine Clock Reaction

Combine iodine, I_2, with iodide, I^-, and they will react to form the triiodide ion, I_3^-, which quickly binds with starch to form a dark blue color when in solution.

$$I_2 \quad + \quad I^- \longrightarrow \quad I_3^-$$

Triiodide-starch complex (dark blue)

Rather than combining I_2 and I^-, you can make a solution of the two by reacting I^- with hydrogen peroxide, H_2O_2, in the presence of hydrogen ions. This reaction is relatively slow but proceeds as follows:

$$\underline{1}\ H_2O_2 + \underline{2}\ I^- + \underline{2}\ H^+ \xrightarrow{slow} \underline{1}\ I_2 + \underline{2}\ H_2O$$

Hydrogen peroxide   iodide ion   hydrogen ion (from an acid)   iodine   water

Balance this equation

1. If you had one hydrogen peroxide molecule combine with two iodine ions with hydrogen ions present, would you ever have I_2 and I^- at the same time? Why or why not?

   Once the I_2 molecule forms, there would be no more I^- ions. So, in this case, you'll not be able to have I_2 and I^- exist at the same time.

2. If you had one trillion hydrogen peroxide molecules combine with two trillion iodine ions with hydrogen ions present, would you ever have I_2 and I^- at the same time? Why or why not?

   Once a few I^- react to form some I_2, there will still be some I^- ions that have yet to react. At this point in time, there will be both I_2 and I^- in solution.

3. How does the slowness of this reaction play a role?

   If this reaction were superfast, then the I_2 and I^- would not coexist for any appreciable time. Because the reaction is relatively slow, however, there's time for the I_2 and the I^- to react to form I_3^-.

4. The reaction of H_2O_2 with I^- to form I_2 is quite slow. How might the rate of this reaction be increased?

   Increase the temperature of the solution. The ions and molecules will be moving faster, which promotes the chemical reaction. You could also increase the concentration of the reactants.

144

Vitamin C causes iodine, I_2, to break down rapidly into iodide ions as per the following reaction:

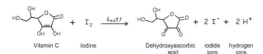

| Vitamin C | Iodine | | Dehydroxyascorbic acid | iodide ions | hydrogen ions |

5. A solution of trillion hydrogen peroxide molecules and two trillion iodide ions (with plenty of hydrogen ions present) forms plenty of iodine, I_2, molecules. But what happens to these I_2 molecules if vitamin C is also in the solution?

The vitamin C "quenches" these iodine molecules the moment they are formed. In chemical terms, the vitamin C is "reacting" with the iodine, I_2, so that this iodine only exists for the briefest moment.

6. Do I_2 and I^- ever coexist, practically speaking, with the vitamin C present? Why or why not?

The I_2 is destroyed by the vitamin C the moment it is formed. So, practically speaking, the I_2 and the I^- are not able to coexist. This means that the triodide ions, I_3^-, are not able to form.

7. As shown in the chemical equation at the top of the page, the vitamin C molecule is no longer a vitamin C molecule after it reacts with I_2. For the above solution, what would eventually happen if only 500 billion vitamin C molecules were present?

As the vitamin C destroys the I_2, so does the I_2 "destroy" the vitamin C. Eventually there would come a point where all of the vitamin C was used up. At that point, the I_2 and the I^- will be able to coexist.

8. At what point will I_2 molecules and I^- ions finally be able to coexist in significant amounts?

Once all of the vitamin C has been consumed in the reaction with iodine, I_2, then the I_2 and I^- will be able to coexist.

9. If starch is also present, what happens to the color of the solution at that point?

The moment the I_2 and I^- are able to coexist is the moment they can react to form I_3^-, which will bind with the starch to give the solution a dark blue color. How long this takes to happen is very reliable and based upon the concentration of the reactants and the temperature of the solution. This is the essense of the popular activity/demo called the "iodine-clock reaction."

10. Would this take a longer or shorter time to happen if the solutions were warmed to a higher temperature? Why?

At a warmer temperature, the transformation to the dark blue color will happen faster as an indication of the faster motion of the ions and molecules.

11. Water disinfected with iodine, I_2, can taste a bit like iodine. Yuk! What might be added to this I_2 disinfected water to remove some of this taste?

Add vitamin C!

145

Chapter 14: Organic Compounds
Structures of Organic Compounds

1. What are the chemical formulas for the following structures?

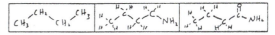

Formula: C_6H_{14} CH_6O C_8H_{18} $C_{10}H_{15}NO$

2. How many covalent bonds is carbon able to form? __4__

3. What is wrong with the structure shown in the box at right?

__THE CARBON OF THE CARBONYL IS__
BONDED 5 TIMES.

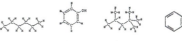

4. a. Draw a hydrocarbon that contains 4 carbon atoms.   b. Redraw your structure and transform it into an amine.   c. Transform your amine into an amide. You may need to relocate the nitrogen.

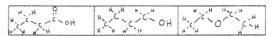

d. Redraw your amide, transforming it into a carboxylic acid.   e. Redraw your carboxylic acid, transforming it into an alcohol.   f. Rearrange the carbons of your alcohol to make an ether.

147

Chapter 14: Organic Compounds
Polymers

1. Circle the monomers that may be useful for forming an addition polymer and draw a box around the ones that may be useful for forming a condensation polymer.

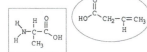

2. Which type of polymer always weighs less than the sum of its parts? Why?

THE CONDENSATION POLYMERS LOSE SMALL MOLECULES, SUCH AS WATER, WHEN THEY FORM AND THUS THE POLYMER THAT FORMS WEIGHS LESS THAN THE SUM OF ITS MONOMERS.

3. Would a material with the following arrangement of polymer molecules have a relatively high or low melting point? Why?

CRYSTALLINE   CRYSTALLINE

WITH ITS MANY CRYSTALLINE REGIONS THIS POLYMER OUGHT TO HAVE A RELATIVELY HIGH MELTING TEMPERATURE.

CRYSTALLINE

149

Chapter 15: The Basic Unit of Life—The Cell
Prokaryotic Cells and Eukaryotic Cells

1. State whether the following features are found in prokaryotic cells, eukaryotic cells, or both.

a. nucleic acids	both
b. cell membrane	both
c. nucleus	eukaryotic cells
d. organelles	eukaryotic cells
e. mitochondria	eukaryotic cells
f. chloroplasts	eukaryotic cells
g. circular chromosome	prokaryotic cells
h. cytoplasm	both

2. State whether these are prokaryotes or eukaryotes.

| Human | Bacteria | Tree |

a. eukaryote   b. prokaryotes   c. eukaryote

| Bird | Mushroom |

d. eukaryote   e. eukaryote

151

305

Conceptual Integrated Science — Third Edition

Chapter 15: The Basic Unit of Life—The Cell
Eukaryotic Organelles

1. Label the organelles in the animal cell with the following terms:

Cytoskeleton   Ribosomes
Golgi apparatus   Rough endoplasmic reticulum
Lysosome   Smooth endoplasmic reticulum
Mitochondrion

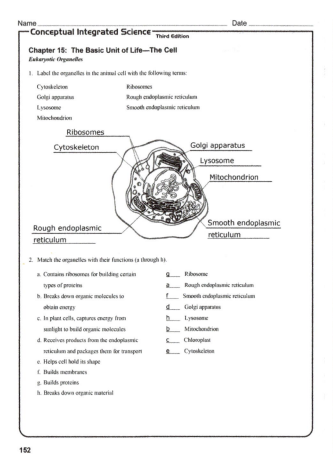

Ribosomes
Cytoskeleton
Golgi apparatus
Lysosome
Mitochondrion
Rough endoplasmic reticulum
Smooth endoplasmic reticulum

2. Match the organelles with their functions (a through h).

a. Contains ribosomes for building certain types of proteins   __g__ Ribosome

b. Breaks down organic molecules to obtain energy   __a__ Rough endoplasmic reticulum

c. In plant cells, captures energy from sunlight to build organic molecules   __f__ Smooth endoplasmic reticulum

d. Receives products from the endoplasmic reticulum and packages them for transport   __d__ Golgi apparatus

e. Helps cell hold its shape   __h__ Lysosome

f. Builds membranes   __b__ Mitochondrion

g. Builds proteins   __c__ Chloroplast

h. Breaks down organic material   __e__ Cytoskeleton

Conceptual Integrated Science — Third Edition

Chapter 15: The Basic Unit of Life—The Cell
The Cell Membrane

1. Label the three components of the cell membrane in the figure.

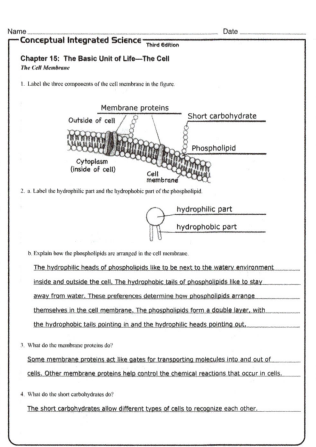

Membrane proteins
Outside of cell
Short carbohydrate
Phospholipid
Cytoplasm (inside of cell)
Cell membrane

2. a. Label the hydrophilic part and the hydrophobic part of the phospholipid.

hydrophilic part
hydrophobic part

b. Explain how the phospholipids are arranged in the cell membrane.

The hydrophilic heads of phospholipids like to be next to the watery environment inside and outside the cell. The hydrophobic tails of phospholipids like to stay away from water. These preferences determine how phospholipids arrange themselves in the cell membrane. The phospholipids form a double layer, with the hydrophobic tails pointing in and the hydrophilic heads pointing out.

3. What do the membrane proteins do?

Some membrane proteins act like gates for transporting molecules into and out of cells. Other membrane proteins help control the chemical reactions that occur in cells.

4. What do the short carbohydrates do?

The short carbohydrates allow different types of cells to recognize each other.

Yearn to learn. What you learn is yours and can never be taken from you.

Conceptual Integrated Science — Third Edition

Chapter 15: The Basic Unit of Life—The Cell
Diffusion and Osmosis

1. The molecules represented by squares move across the cell membrane through diffusion in the diagram on the right.

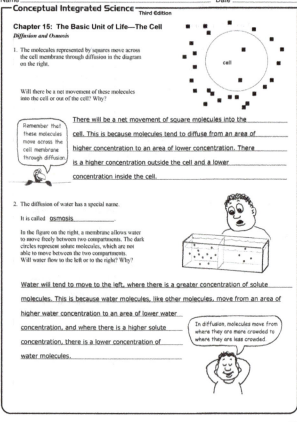

cell

Will there be a net movement of these molecules into the cell or out of the cell? Why?

Remember that these molecules move across the cell membrane through diffusion.

There will be a net movement of square molecules into the cell. This is because molecules tend to diffuse from an area of higher concentration to an area of lower concentration. There is a higher concentration outside the cell and a lower concentration inside the cell.

2. The diffusion of water has a special name.

It is called __osmosis__ .

In the figure on the right, a membrane allows water to move freely between two compartments. The dark circles represent solute molecules, which are not able to move between the two compartments. Will water flow to the left or to the right? Why?

Water will tend to move to the left, where there is a greater concentration of solute molecules. This is because water molecules, like other molecules, move from an area of higher water concentration to an area of lower water concentration, and where there is a higher solute concentration, there is a lower concentration of water molecules.

In diffusion, molecules move from where they are more crowded to where they are less crowded.

Name _____ Date _____

Conceptual Integrated Science — Third Edition

Chapter 15: The Basic Unit of Life—The Cell
Facilitated Diffusion and Active Transport

1. Which of these describe facilitated diffusion, and which describe active transport?

   a. Does not require energy from the cell __Facilitated diffusion__

   b. Requires energy from the cell __Active transport__

   c. Moves molecules from a region of low concentration to a region of high concentration

   __Active transport__

   d. Moves molecules from a region of high concentration to a region of low concentration

   __Facilitated diffusion__

2. Which of the following shows facilitated diffusion, and which shows active transport?

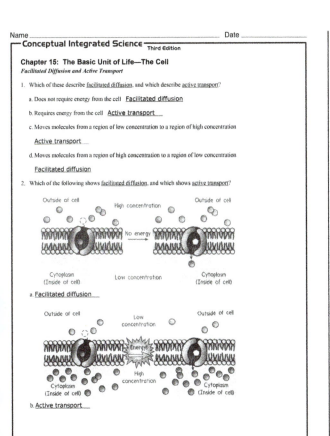

   a. __Facilitated diffusion__

   b. __Active transport__

156

Name _____ Date _____

Conceptual Integrated Science — Third Edition

Chapter 15: The Basic Unit of Life—The Cell
Cell Division

1. What is the function of cell division for single-celled organisms?

   __Cell division allows single-celled organisms to reproduce.__

2. What are some functions of cell division for multicellular organisms?

   __Multicellular organisms use cell division to develop, grow, and maintain their tissues.__

3. What happens during each stage of the cell cycle?

   Gap 1 __The cell doubles in size, doubling its mitochondria and other organelles.__

   Synthesis __The cell makes a copy of its DNA.__

   Gap 2 __The cell builds the machinery needed for cell division.__

   Mitosis and cytokinesis __The cell divides and in cytokinesis the cytoplasm divides.__

4. Mitosis is divided into a number of phases. Describe what happens during each of the phases listed below.

   Prophase __The chromosomes condense and the nuclear membrane breaks down.__

   Metaphase __The chromosomes are lined up at the equatorial plane.__

   Anaphase __The sister chromatids are pulled apart.__

   Telophase __The chromosomes decondense and new nuclear membranes form.__

157

Name _____ Date _____

Conceptual Integrated Science — Third Edition

Chapter 15: The Basic Unit of Life—The Cell
Photosynthesis

1. During photosynthesis, one kind of energy, __light energy from the Sun,__
   is converted into __chemical energy in organic molecules__

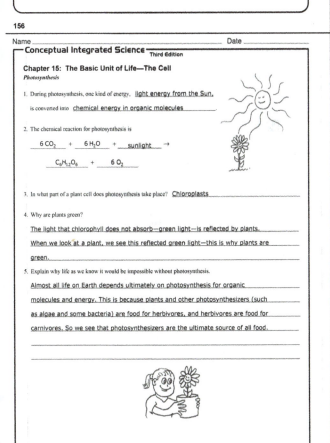

2. The chemical reaction for photosynthesis is

   $6\ CO_2$ + $6\ H_2O$ + __sunlight__ →

   $C_6H_{12}O_6$ + $6\ O_2$

3. In what part of a plant cell does photosynthesis take place? __Chloroplasts__

4. Why are plants green?

   __The light that chlorophyll does not absorb—green light—is reflected by plants.__
   __When we look at a plant, we see this reflected green light—this is why plants are__
   __green.__

5. Explain why life as we know it would be impossible without photosynthesis.

   __Almost all life on Earth depends ultimately on photosynthesis for organic__
   __molecules and energy. This is because plants and other photosynthesizers (such__
   __as algae and some bacteria) are food for herbivores, and herbivores are food for__
   __carnivores. So we see that photosynthesizers are the ultimate source of all food.__

158

Name _____ Date _____

Conceptual Integrated Science — Third Edition

Chapter 15: The Basic Unit of Life—The Cell
Cellular Respiration

1. The chemical reaction for cellular respiration is

   $C_6H_{12}O_6$ + $6O_2$ + about 38 ADP →

   $6CO_2$ + $6H_2O$ + about 38 ADP

2. Cells use cellular respiration to produce ATP. How do cells obtain energy from ATP?

   __Energy is obtained from ATP when one of its three phosphate groups is removed,__
   __leaving ADP.__

Circle the correct answers.
3. Which of the processes requires oxygen?

   (Glycolysis)

   (Krebs cycle and electron transport)

   (Alcoholic fermentation)

   (Lactic acid fermentation)

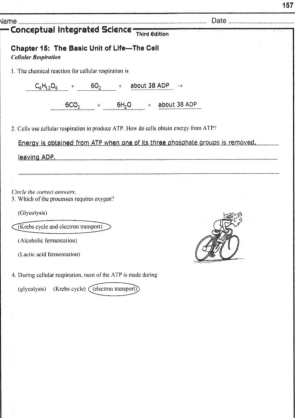

4. During cellular respiration, most of the ATP is made during

   (glycolysis) (Krebs cycle) (electron transport)

159

307

Conceptual Integrated Science — *Third Edition*

Chapter 16: Genetics
DNA Replication

1. The following is a piece of DNA:

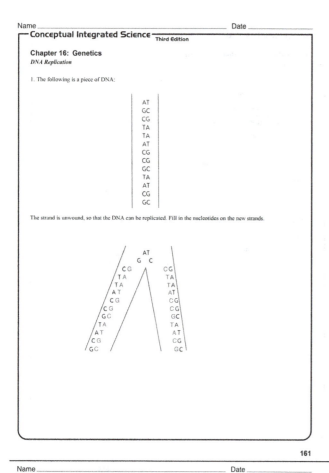

```
AT
GC
CG
TA
TA
AT
CG
CG
GC
TA
AT
CG
GC
```

The strand is unwound, so that the DNA can be replicated. Fill in the nucleotides on the new strands.

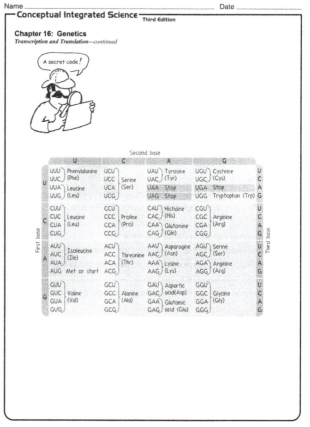

Jean's genes are mostly what make Jean Jean.

Conceptual Integrated Science — *Third Edition*

Chapter 16: Genetics
Transcription and Translation

1. The figure below shows how information from DNA is used to build a protein. Write the names of the appropriate processes above the arrows.

DNA → **Transcription** → RNA → **Translation** → Protein

2. Transcription takes place in the cell's __nucleus__.

During transcription, DNA is used to make a molecule of __messenger RNA or mRNA__.

3. If the following strand of DNA is transcribed, what are the nucleotides found on the transcript?

__U A C C A G U A U G C A U G U U A C (mRNA transcript)__

A T G G T C A T A C G T A C A A T G

4. Translation takes place in the cell's __cytoplasm__. Translation is performed by

organelles called __ribosomes__.

5. Divide the transcript from your answer to Question 3 into codons. Then, figure out the sequence of amino acids assembled in the ribosome. Use the genetic code table on the next page.

Transcript __UAC CAG UAU GCA UGU UAC__

Codons __UAC-CAG-UAU-GCA-UGU-UAC__

Amino acids __Tyr-Gln-Tyr-Ala-Cys-Tyr__

Conceptual Integrated Science — *Third Edition*

Chapter 16: Genetics
Transcription and Translation—continued

A secret code!

Second base

First base	U	C	A	G	Third base
U	UUU UUC Phenylalanine (Phe) UUA UUG Leucine (Leu)	UCU UCC UCA UCG Serine (Ser)	UAU UAC Tyrosine (Tyr) UAA Stop UAG Stop	UGU UGC Cysteine (Cys) UGA Stop UGG Tryptophan (Trp)	U C A G
C	CUU CUC CUA CUG Leucine (Leu)	CCU CCC CCA CCG Proline (Pro)	CAU CAC Histidine (His) CAA CAG Glutamine (Gln)	CGU CGC CGA CGG Arginine (Arg)	U C A G
A	AUU AUC AUA Isoleucine (Ile) AUG Met or start	ACU ACC ACA ACG Threonine (Thr)	AAU AAC Asparagine (Asn) AAA AAG Lysine (Lys)	AGU AGC Serine (Ser) AGA AGG Arginine (Arg)	U C A G
G	GUU GUC GUA GUG Valine (Val)	GCU GCC GCA GCG Alanine (Ala)	GAU GAC Aspartic acid (Asp) GAA GAG Glutamic acid (Glu)	GGU GGC GGA GGG Glycine (Gly)	U C A G

Conceptual Integrated Science
Third Edition

Chapter 16: Genetics
Genetic Mutations

1. Define the following terms:

Genetic mutation _____ a change in the sequence of nucleotides in an

organism's DNA _____

Point mutation _____ the substitution of one nucleotide for another _____

Nonsense mutation _____ a mutation that produces a stop codon in the

middle of a protein coding sequence _____

Frameshift mutation _____ the insertion or deletion of nucleotides that

causes the sequence of amino acids in a protein to be completely changed _____

2. Translate the following mRNA sequence into amino acids. You can use the genetic code table on the next page.

AAU	GUC	CCG	ACC	AAA	GCU
asparagine	valine	proline	threonine	lysine	alanine

3. What point mutation in the sequence above could cause the substitution of the amino acid serine for asparagine?

A change from AAU to AGU in which a point mutation changes the second "A" to "G"

would result in a serine amino acid instead of asparagine.

4. How could a change in a single nucleotide in the sequence above result in a nonsense mutation?

A change from AAA to UAA in which the first "A" becomes a "U" would produce a

stop codon and thus a nonsense mutation.

5. The insertion or deletion of one or two nucleotides causes a frameshift mutation. Why doesn't the insertion or deletion of three nucleotides cause a frameshift mutation as well?

The insertion of three nucleotides causes the insertion of an amino acid (and may

change a few amino acids as well, depending on where the insertion occurs). It does

not *completely* change the sequence of amino acids in the protein the way a

frameshift mutation does. The same holds true for the deletion of three nucleotides.

This is because a codon consists of three nucleotides.

Conceptual Integrated Science
Third Edition

Second base

	U	C	A	G	
U	UUU } Phenylalanine (Phe) UUC	UCU UCC } Serine (Ser) UCA UCG	UAU } Tyrosine (Tyr) UAC UAA Stop UAG Stop	UGU } Cysteine (Cys) UGC UGA Stop UGG Tryptophon (Trp)	U C A G
	UUA } Leucine (Leu) UUG				
C	CUU CUC } Leucine (Leu) CUA CUG	CCU CCC } Proline (Pro) CCA CCG	CAU } Histidine (His) CAC CAA } Glutamine (Gln) CAG	CGU CGC } Arginine (Arg) CGA CGG	U C A G
A	AUU AUC } Isoleucine (Ile) AUA AUG Met or start	ACU ACC } Threonine (Thr) ACA ACG	AAU } Asparagine (Asn) AAC AAA } Lysine (Lys) AAG	AGU } Serine (Ser) AGC AGA } Arginine (Arg) AGG	U C A G
G	GUU GUC } Valine (Val) GUA GUG	GCU GCC } Alanine (Ala) GCA GCG	GAU } Aspartic acid(Asp) GAC GAA } Glutamic acid (Glu) GAG	GGU GGC } Glycine (Gly) GGA GGG	U C A G

First base / Third base

Conceptual Integrated Science
Third Edition

Chapter 16: Genetics
Chromosomes

1. On the left are some chromosomes from a diploid cell. On the right are chromosomes from a human cell.

a. How many chromosomes are in the diploid cell? **4**

b. After the diploid cell goes through meiosis, how many chromosomes will there be in the resulting cells? **2**

c. How many chromosomes are in the human cell? **46**

d. Two kinds of cells that are produced through meiosis in humans are

eggs and **sperm**.

e. The human chromosomes above belong to one of the two people shown here. Circle the correct person.

f. How do you know?

There is an X and a Y chromosome, so the chromosomes must be from the man.

Conceptual Integrated Science
Third Edition

Chapter 16: Genetics
Comparing Mitosis and Meiosis

1. Give some examples of when organisms use mitosis.

Single-celled organisms use mitosis to reproduce. Multicellular organisms use

mitosis to develop and grow, as well as to maintain their bodies.

2. a. When do organisms use meiosis?

Meiosis is used to make haploid cells.

b. What kinds of cells are produced through meiosis?

Haploid cells such as eggs and sperm.

Circle the correct answers.

3. a. In meiosis, the number of cells produced is (one) (two) (three) **(four)**

In mitosis, the number of cells produced is (one) **(two)** (three) (four).

b. During meiosis, cells produced are **(haploid)** (diploid).

During mitosis, cells produced are (haploid) **(diploid)**.

c. The cells produced are **(different from one another)** (identical) in meiosis.

The cells produced are (different from one another) **(identical)** in mitosis.

d. Crossing over happens during (mitosis) **(meiosis)**.

Conceptual Integrated Science — Third Edition

Chapter 16: Genetics
Dominant and Recessive Traits

1. Some small woodland creatures have either spots or stripes. Fur pattern is determined by a single gene. The striped phenotype is **dominant** and the spotted phenotype is **recessive**.

   Both the woodland creatures shown below are **homozygotes**.

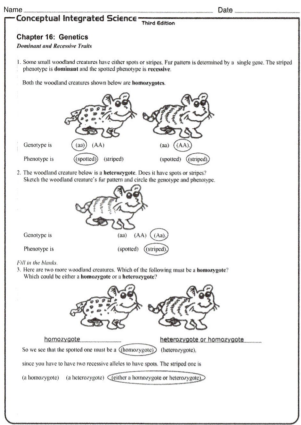

   Genotype is   (aa) (AA)   (aa) (AA)
   Phenotype is   (spotted) (striped) (spotted) (striped)

2. The woodland creature below is a **heterozygote**. Does it have spots or stripes? Sketch the woodland creature's fur pattern and circle the genotype and phenotype.

   Genotype is   (aa)   (AA)   (Aa)
   Phenotype is   (spotted) (striped)

Fill in the blanks.

3. Here are two more woodland creatures. Which of the following must be a **homozygote**? Which could be either a **homozygote** or a **heterozygote**?

   homozygote   heterozygote or homozygote

   So we see that the spotted one must be a (homozygote) (heterozygote),

   since you have to have two recessive alleles to have spots. The striped one is

   (a homozygote)   (a heterozygote)   (either a homozygote or heterozygote).

169

Conceptual Integrated Science — Third Edition

Chapter 16: Genetics
Inheritance Patterns

1. Suppose you breed two woodland creatures together. One individual has genotype aa, and the other has genotype AA. Draw the cross below.

   Genotype is   (aa) (AA)   (aa) (AA)
   Phenotype is   (spotted) (striped) (spotted) (striped)

2. What are the progeny like?

   Genotype is   (aa)   (AA)   (Aa) and phenotype is (spotted)   (striped).

Fill in the table below.

3. Now you breed two of the progeny from Question 2 together. What will their offspring look like?

Allele received from father

	A	a
Allele received from mother A	Genotype: AA Phenotype: striped	Genotype: Aa Phenotype: striped
a	Genotype: Aa Phenotype: striped	Genotype: aa Phenotype: spotted

So, the offspring include woodland creatures that are

(spotted)   (striped)   (both spotted and striped).

The offspring phenotypes are found in a ratio of **3** striped: **1** spotted.

170

Conceptual Integrated Science — Third Edition

Chapter 16: Genetics
Genetic Engineering

1. What is genetic engineering?

   Genetic engineering is the process of changing the traits of organisms by directly

   manipulating their DNA.

2. Carla learns that some people have a condition in which their bodies do not make enough human growth hormone. Because human growth hormone promotes growth, making too little of the hormone results in unusually short stature. Doctors give these patients injections of human growth hormone in order to help them achieve normal heights.

   Carla asks, "But how do the doctors get human growth hormone for the injections? They can't take hormone from other people, can they?"

   You say, "No, they don't collect it from other people. They use genetically engineered organisms that produce human growth hormone." Explain to Carla how this works.

   Scientists insert the gene for human growth hormone into the DNA of microorganisms

   such as the bacteria *E. coli*. The bacteria then make human growth hormone from the

   gene. This hormone is then collected and purified before being used in medicine.

3. Dirk wants to know why some people are against using genetically engineered organisms in food and agriculture. Explain at least two potential issues.

   Some scientists worry about the effect of genetically engineered organisms on natural

   habitats and ecosystems. Some genetically engineered crops encourage the use of more

   chemical pesticides. Selection for resistant plants may also cause the evolution of

   superweeds. The genes of GM crops could spread into natural populations. Finally,

   genetically modified seeds may be more expensive for farmers without providing

   benefits.

171

Conceptual Integrated Science — Third Edition

Chapter 17: The Evolution of Life
Genetic Variation in Human Traits

1. Do the human traits listed below show genetic variation?

   a. age   no
   b. eye color   yes
   c. number of toes   no
   d. curliness or straightness of hair   yes
   e. presence or absence of dimples   yes
   f. upright posture   no
   g. owning versus not owning a dog   no
   h. height   yes

2. What kinds of traits have the potential of evolving via natural selection?

   Traits that show genetic variation. In other words, inherited traits that vary from individual

   to individual.

173

Conceptual Integrated Science — Third Edition

Chapter 17: The Evolution of Life
Natural Selection

1. On a tropical island, there lives a species of bird called the Sneezlee bird. Sneezlee birds eat seeds and also show genetic variation in beak size. Some Sneezlee birds have bigger beaks, and some have smaller beaks.

a. Draw a box around the bird with the biggest beak.

b. Put a check mark by the bird with the smallest beak.

2. One year, there are not many seeds available for Sneezlee birds due to a drought in the summer. Small seeds are quickly eaten by the Sneezlee birds. Only larger, tougher seeds are left. Sneezlee birds with larger, stronger beaks are better at cracking these larger seeds.

a. Draw a box around the two birds most likely to survive the drought.

b. Mark X's through two Sneezlee birds that are more likely to die.

3. Beak size is an inherited trait. Parent birds with larger, stronger beaks tend to have offspring with larger, stronger beaks. How would the Sneezlee bird population evolve due to the drought?

On average, the Sneezlee birds after the drought have larger, stronger beaks.

Conceptual Integrated Science — Third Edition

Chapter 17: The Evolution of Life
Size and Shape

1. The imaginary mammal below occupies temperate forests in the Eastern United States.

a. If a population of these mammals moved and successfully colonized an Arctic habitat, how might you predict that it would evolve? Draw the Arctic form of the mammals below.

Think about ear and leg length

b. If a population of these mammals moved and successfully colonized a desert habitat, how might you predict that it would evolve? Draw the desert form of the mammals below.

c. Explain your drawings.

The heat a mammal loses depends on its surface area, because heat is lost through the body's surface. In a cold environment such as the Arctic, mammals evolve smaller ears and shorter legs. These features decrease surface area and help conserve heat. In a hot desert, mammals evolve larger ears and longer legs. These features increase surface area and help the animals lose heat.

Think area vs volume.

Conceptual Integrated Science — Third Edition

Chapter 17: The Evolution of Life
Mechanisms of Evolution

Suppose you have a population of peppered moths in which the allele frequencies are 50% light allele and 50% dark allele. For each of the events below, list the mechanism of evolution involved and the event's effect on allele frequencies.

Event	Mechanism of evolution (natural selection, mutation pressure, genetic drift, or gene flow)	Effect on allele frequencies (increases frequency of the light allele, increases frequency of the dark allele, or change cannot be predicted)
1. Some light moths migrate into the population from a nearby unpolluted area.	Gene flow	Increases frequency of the light allele
2. Pollution in the town increases.	Natural selection	Increases frequency of the dark allele
3. A storm kills half the moths in the population.	Genetic drift	Change cannot be predicted
4. Some dark moths migrate into the population from a nearby polluted area.	Gene flow	Increases frequency of the dark allele
5. Just by chance, dark moths leave more offspring than light moths one year.	Genetic drift	Increases frequency of the dark allele
6. Dark moths survive better than light moths.	Natural selection	Increases frequency of the dark allele

---

Conceptual Integrated Science — Third Edition

Chapter 17: The Evolution of Life
Speciation

1. Define the following terms:

Species — A group of organisms whose members can breed with one another but not with members of other species

Speciation — The formation of a new species

Reproductive barrier — A feature that prevents two groups of organisms from interbreeding

2. Explain the difference between a prezygotic reproductive barrier and a postzygotic reproductive barrier. Then, give one example of each.

A prezygotic reproductive barrier is one that acts before fertilization. A postzygotic reproductive barrier is one that acts after fertilization. An example of a prezygotic reproductive barrier is mating at a different time of year or having different courtship rituals. An example of a postzygotic reproductive barrier is having hybrid offspring that don't survive or are sterile. (Note that other examples are possible.)

3. A geographic barrier –a newly formed river– divides a population of frogs into two separate populations. Explain how this event could result in speciation.

After the river divides the population into two separate populations, each population evolves independently from the other. The populations are likely to evolve differences as each adapts to its own environment. Over time, this could result in the evolution of a reproductive barrier that would make the two populations separate species.

┌─ Conceptual Integrated Science ─ Third Edition

Chapter 18: Diversity of Life on Earth
Classification

Circle the correct answer:
1. Linnaean classification involves grouping species together on the basis of how
   (**similar**) (different) they are.

2. Fill in the levels of Linnaean classification from the largest group to the smallest group.

   Domain

   Kingdom

   Phylum

   Class

   Order

   Family

   Genus

   Species

Circle the correct answers:
3. The scientific name of a species consists of its (**genus**) (family) (species) name

   and its (genus) (family) (**species**) name.

> Earth is home to as many as 10 million different species! Humans are just one of these.

┌─ Conceptual Integrated Science ─ Third Edition

Chapter 18: Diversity of Life on Earth
Evolutionary Trees

Circle the correct answers:
1. Biologists now try to classify organisms on the basis of how (**similar**) (closely related)
   they are to each other.

2. The evolutionary tree below shows how three species—the daisy, the honey mushroom, and the gentoo penguin—are related.

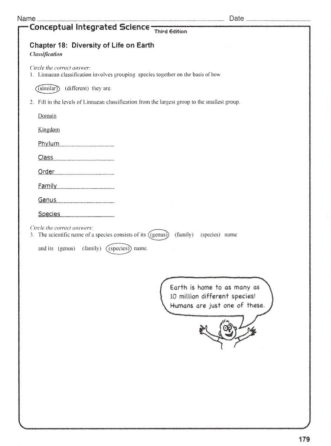

   daisy   honey mushroom   gentoo penguin

   a. This evolutionary tree tells us that the (daisy) (**honey mushroom**) (gentoo penguin)

      and (daisy) (honey mushroom) (**gentoo penguin**) are more closely related to each

      other than either is to the (**daisy**) (honey mushroom) (gentoo penguin).

   b. Place a dot at the point where the lineage that eventually gave rise to daisies split from the lineage that eventually gave rise to gentoo penguins.

   c. Now place an asterisk at the point where the lineage that eventually gave rise to honey mushrooms split from the lineage that eventually gave rise to gentoo penguins.

┌─ Conceptual Integrated Science ─ Third Edition

Chapter 18: Diversity of Life on Earth
Bacteria and Archaea

Fill in the blanks:
1. The three domains of life are

   Bacteria_____, Archaea_____, and Eukarya_____.

   Of the 3 domains, Bacteria_____ and Archaea_____ consist of

   prokaryotes and Eukarya_____ consists of eukaryotes.

Circle the correct answers:
2. a. Are there any bacteria that can make their own food through photosynthesis, the way plants do? (**Yes**) (No)

   b. Are there any bacteria that get their food from other organisms, the way animals do? (**Yes**) (No)

3. Bacteria reproduce (sexually) (**asexually**).

4. Bacteria that live on our bodies benefit us by

   (producing vitamins)

   (keeping dangerous bacteria from invading our bodies)

   (**both of these**).

5. Are all archaea extremophiles? (Yes) (**No**)

6. Some chemoautotrophs make food using

   (**chemical energy**)

   (energy from sunlight).

> In life, it's the stuff you can't see that does us the most damage.

┌─ Conceptual Integrated Science ─ Third Edition

Chapter 18: Diversity of Life on Earth
Domain Eukarya and Protists

1. All the living things found in the domain Eukarya have (prokaryotic cells) (**eukaryotic cells**).

2. Name the four kingdoms that make up the domain Eukarya.

   Protists

   Plants

   Fungi

   Animals

3. Protists get food from photosynthesis or from other organisms. Place the protists below into the correct column.

   Amoebas

   Kelp

   Diatoms

   Ciliates

   Plasmodium, the protist that causes malaria

> Kelp is a protist!

Get food from photosynthesis	Get food from other organisms
Diatoms	Amoebas
Kelp	Ciliates
_____	Plasmodium
_____	_____

Chapter 18: Diversity of Life on Earth
Plants

1. Match the following plant structures with their function.

Roots	c_____	a. Distributes water and other resources
Shoots	b_____	b. Conduct photosynthesis
Vascular system	a_____	c. absorb water and nutrients from soil

Circle the correct answers:
2. Mosses (do) (do not) have vascular systems whereas ferns
   (do) (do not) have vascular systems.

3. (Mosses) (Ferns) (Mosses and ferns) (Seed plants) have to live in a moist
   environment because sperm must swim through the environment to fertilize eggs.

4. Describe each structure and write the name of the group of plants that possess each of
   the structures.

Structure	Description	Plant group in which structure is found
Pollen	male reproductive cells wrapped in a protective coating	seed plants
Seed	plant embryo and food supply wrapped in a tough outer coating	seed plants
Flower	reproductive structure of flowering plants	flowering plants
Fruit	structure surrounding seeds that helps spread the seeds around	flowering plants
Cone	reproductive structure of conifers	conifers

Chapter 18: Diversity of Life on Earth
Animals

Match the following animal groups with the list of features on the right. Note that some groups
have more than one answer!

Sponges	k_____	a. Clams, oysters, and squids are all part of this group.
Cnidarians	f, n_____	b. The swim bladder of these animals makes their density the same as the density of water—this makes it much easier for them to swim well!
Flatworms	v_____	c. These animals sink if they stop swimming.
Roundworms	e_____	d. Many of the animals in this group start life in the water and then move to land as adults.
Arthropods	i, p, s_____	e. All their muscles run from head to tail. Because of this, these animals move like flailing whips.
Mollusks	a, x_____	f. There is a sedentary polyp stage and a swimming medusa stage.
Annelids	g, h_____	g. Leeches belong to this group.
Echinoderms	m, r_____	h. This group includes worms whose bodies are divided into segments.
Chordates	q_____	i. Their bodies are divided into segments and their legs have bendable joints.
Cartilaginous fishes	c, t_____	j. Their skins are made of living cells that can dry out, so they have to stay in moist environments.
Ray-finned fishes	b_____	k. In these animals, a constant flow of water comes in through many pores and goes out the top. The purpose of this constant flow is to catch food.
Amphibians	j, z, d_____	l. Birds and crocodiles are examples of this group of animals.
Reptiles	l_____	m. These animals move using tube feet.
Birds	o_____	n. Watch out! Their tentacles have barbed stinging cells.
Mammals	u, w_____	o. They have wings and hollow bones.
		p. The insects are part of this group.
		q. The vertebrates are part of this group.
		r. Starfish belong to this group.
		s. These animals shed their exoskeleton as they grow.
		t. These animals, such as sharks and rays, have a skeleton made of cartilage.
		u. These animals have hair and feed their young milk.
		v. Tapeworms -parasites that live in the intestines of humans and other animals- belong to this group.
		w. The platypus belongs to this group, and so do bats and humans!
		x. Most of the animals in this group have a shell, although slugs and octopuses don't.
		y. They are flying endotherms.
		z. Frogs belong to this group.

Chapter 18: Diversity of Life on Earth
Viruses and Prions

1. Is a virus a prokaryote, a eukaryote, or neither? Explain your answer.

   A virus is not a cell, and it is not made of cells, so it is neither a prokaryote nor a
   eukaryote.

2. Describe the structure of a virus.

   A virus is a small piece of genetic material wrapped in a protein coat. A virus's genetic
   material may be made of DNA (either double stranded or single stranded) or RNA
   (again, either double stranded or single stranded).

3. How do viruses "reproduce"?

   Viruses infect a host cell. They then use the host cell's machinery and resources to copy
   their own viral genetic material and to build viral proteins for the protein coat. The
   genetic material and protein coat are assembled to form new viruses, which then leave
   the host cell to infect other cells.

4. What is a prion? How does a prion cause disease?

   Prions are proteins that are incorrectly folded. Prions cause disease by infecting cells
   and then converting properly folded proteins to the incorrectly folded form. They are
   responsible for mad cow disease and Creutzfeld-Jacob disease, both of which are
   associated with brain damage.

Chapter 19: Human Biology I—Control and Development
Parts of the Brain

Match the parts of the brain with their body functions.
Note that some parts have more than one function.

Brainstem	d_____	a. Deals with visual information (what we see)
Cerebellum	b, k_____	b. Controls balance, posture, and coordination
Cerebrum	a, c, e, f, i_____	c. Deals with sensory information about temperature, touch, and pain
Thalamus	g_____	d. Controls basic involuntary activities such as heartbeat, respiration, and digestion
Hypothalamus	h, j_____	e. Allows us to understand spoken language
		f. Controls our voluntary movements
		g. Sorts and filters information and then passes it to the cerebrum
		h. Responsible for emotions such as pleasure and rage
		i. Controls our speech
		j. Controls hunger, thirst, and sex drive
		k. Controls the fine movements we use in activities that we perform "without thinking"

313

Chapter 19: Human Biology I—Control and Development
The Nervous System

1. The two main parts of the nervous system are the central nervous system

   and peripheral nervous system .

   The central nervous system consists of the brain   and the

   spinal cord .

2. The three types of neurons are sensory neurons

   interneurons , and motor neurons

   Messages from the senses to the central nervous are carried by

   sensory neurons . Neurons that connect one neuron to another neuron are

   interneurons . Messages are carried from the central nervous system

   to muscle cells or to other responsive organs by motor neurons .

3. Motor neurons are further divided into two groups:

   the somatic nervous system .

   which controls voluntary actions and stimulates our voluntary muscles,

   and the autonomic nervous system .

   which controls involuntary actions and stimulates our involuntary muscles

   and other internal organs.

   The autonomic nervous system includes a sympathetic division

   that promotes a "fight or flight" response and a parasympathetic division

   that operates in times of relaxation.

Chapter 19: Human Biology I—Control and Development
Parts of a Neuron

1. a. Label the parts of the neuron in the diagram.

   Dendrites

   Cell body

   Axon

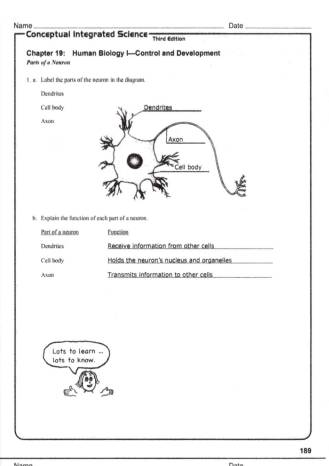

   b. Explain the function of each part of a neuron.

Part of a neuron	Function
Dendrites	Receive information from other cells
Cell body	Holds the neuron's nucleus and organelles
Axon	Transmits information to other cells

   Lots to learn ...
   lots to know.

Chapter 19: Human Biology I—Control and Development
Action Potentials

1. This is a neuron at rest. The neuron is at its resting potential. Draw a + sign on the side of the membrane that is positively charged. Draw a – sign on the side of the membrane that is negatively charged.

   Sodium channel   Potassium channel

   Inside   Axon
   neuron

   Outside neuron

2. Now the neuron fires! There is an action potential. The sodium channels open. Use arrows to show how the sodium ions move. Draw a + sign on the side of the membrane that is positively charged. Draw a – sign on the side of the membrane that is negatively charged.

   Na^+

   Na^+

3. Now the sodium channels close, and the potassium channels open. Use arrows to show how the potassium ions move. Draw a + sign on the side of the membrane that is positively charged. Draw a – sign on the side of the membrane that is negatively charged. The action potential is over.

   K^+ K^+

Chapter 19: Human Biology I—Control and Development
Senses

1. The light-sensitive cells are found in the part of the eye

   called the retina .

   The two types of light-sensitive cells are rods

   and cones .

2. State whether the following describe rods or cones.

   a. Vision at night or in dim light   rods

   b. Let us see color   cones

   c. Detect only black, white, and shades of gray   rods

   d. Not very good at making out fine details   rods

   e. Nonfunctioning version of these causes colorblindness   cones

   Number from 1 to 4:

3. The ear consists of 3 parts: the outer, middle, and inner ear. Sound moves through the air into the ear in the following order:

   3 ____ middle ear bones

   1 ____ pinna

   4 ____ cochlea

   2 ____ eardrum

4. List the five basic tastes.

   sweet   salty   sour

   bitter   umami

┌─ Conceptual Integrated Science ─── Third Edition

Chapter 19: Human Biology I—Control and Development
Hormones

1. What is a hormone?

   A hormone is a molecule that gives instructions to the body. It is produced in one place in the body, released into the blood, and received by target cells elsewhere in the body.

2. What are the two types of hormones? List 2 differences between them.

   The two types of hormones are protein hormones and steroid hormones. Protein hormones bind to receptors on the cell membrane, whereas steroid hormones bind to receptors inside the cell. Protein hormones start a series of chemical reactions in the cell that result in the cell's response, whereas steroid hormones (and their receptors) enter the cell nucleus and directly affect gene transcription.

3. Explain the function of each of the following hormones.

Insulin	Decreases blood glucose levels
Melatonin	Regulates the body's internal clock
Antidiuretic hormone	Causes more water to be retained in the body
Epinephrine	Promotes the body's "fight or flight" response
Parathyroid hormone	Increases blood calcium levels
Oxytocin	Stimulates contraction of the uterus

4. Explain how hormones are involved in maintaining homeostasis in the body.

   Many hormones are found in pairs with opposing effects. For example, insulin decreases blood glucose levels whereas glucagon increases blood glucose levels. These hormones are used to regulate blood glucose levels, increasing it when it is too low (glucagon) and decreasing it when it is too high (insulin).

┌─ Conceptual Integrated Science ─── Third Edition

Chapter 19: Human Biology I—Control and Development
Skeleton

1. List three functions of the skeleton.

   Protects the body

   Supports the body

   Helps to move the body (along with the muscles)

2. Label the three layers of bones.

   Bone marrow

   Compact bone

   Spongy bone

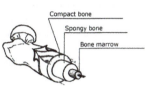

Compact bone
Spongy bone
Bone marrow

3. What is the function of red bone marrow?

   Red bone marrow makes red and white blood cells.

No bones about it!

┌─ Conceptual Integrated Science ─── Third Edition

Chapter 19: Human Biology I—Control and Development
Muscle Contraction

1. Muscles work by (lengthening) (shortening).

2. Two kinds of proteins are involved in muscle contraction:

   actin _____ and

   myosin _____.

3. Draw the myosin heads (ovals) during muscle contraction in the diagram below.

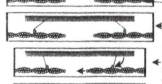

— Myosin
— Myosin head
— Actin

◄ Myosin heads bind to actin.

◄ Myosin heads pivot, causing myosin and actin fibers to slide relative to one another and sarcomere to shorten.

◄ Myosin heads release.

◄ Myosin heads reattach.

◄ Myosin heads pivot again, causing further muscle contraction.

Life is what happens to you while you're making other plans.

Conceptual Integrated Science *Third Edition*

Chapter 20: Human Biology II—Care and Maintenance
Integration of Body Systems

1. The circulatory system and respiratory system work together to help keep the body's tissues supplied with oxygen. Explain how each of the two systems contributes.

    The respiratory system moves air in and out of the lungs, so that the amount of oxygen

    in the alveoli remains high. The circulatory system collects oxygen from the alveoli and

    then carries it to tissues throughout the body.

2. Which body systems are involved in the task you are performing now, reading this question and writing your response?

    The eyes take in light information from the page. The brain (nervous system) processes

    this sensory information and makes sense of what the marks on the page mean. The

    brain also formulates the response and instructs the skeleton and muscles on how to

    move to write the response. The skeleton and muscles move the pen to write your

    answer.

3. Choose another activity that you perform regularly—such as eating a meal, going swimming, brushing your hair, or another activity of your choice—and explain how multiple body systems contribute to that activity.

    Many answers are possible. To take the example of eating a meal, this activity involves

    your skeleton and muscles (putting food in your mouth and chewing it), your senses

    (tasting and smelling the food), and your digestive system (providing saliva to moisten

    and break down food, swallowing, digesting). The nervous system controls all these

    activities.

Conceptual Integrated Science *Third Edition*

Chapter 20: Human Biology II—Care and Maintenance
Circulatory System

Fill in the blanks:

1. Each heartbeat begins in a part of the right atrium called the sinoatrial node _____, or pacemaker.

    The pacemaker starts an action potential that sweeps quickly through two chambers of the heart, the right atrium _____ and left atrium _____, which contract simultaneously.

    The signal also travels to the atrioventricular node _____, and from there to the other two chambers of the heart, the right ventricle _____ and left ventricle _____. These two chambers also contract simultaneously.

2. a. The sound of a heartbeat is "lub-dubb, lub-dubb." What is the "lub"?

    The "lub" is the sound of valves between the atria and ventricles snapping shut

    after the two atria contract.

    b. What is the "dubb"?

    The "dubb" is the sound of the valves between the ventricles and blood vessels

    snapping shut after the contraction of the ventricles.

3. The three types of cells found in blood are red blood cells _____,

    white blood cells _____, and platelets _____.

    Red blood cells _____ transport oxygen to the body's tissues.

    White blood cells _____ are part of the immune system and help our bodies

    defend against disease. Platelets _____ are involved in blood clotting.

4. The molecule in red blood cells that carries oxygen is called hemoglobin _____.

Conceptual Integrated Science *Third Edition*

Chapter 20: Human Biology II—Care and Maintenance
Circulatory System—continued

Number from 1 to 9:
5. In what order does blood flow around the body? Begin with the right atrium.

3	Arteries to lungs
1	Right atrium
5	Left atrium
9	Veins from body tissues
4	Veins from lungs
8	Capillaries near body tissues
7	Arteries to body tissues
2	Right ventricle
6	Left ventricle

Don't let what you can't do interfere with what you can do.

Conceptual Integrated Science *Third Edition*

Chapter 20: Human Biology II—Care and Maintenance
Respiratory System

1. Match each part of the respiratory system with its description.

Nasal passages	e	a.	Where gas exchange occurs
Larynx	c	b.	Another word for trachea
Trachea	h	c.	Structure allows us to speak
Alveoli	a	d.	Raises the ribcage when we inhale
Diaphragm	f	e.	Smelling happens here
Rib muscles	d	f.	Dome-shaped muscle helps us inhale
Bronchi	g	g.	Tubes that go to the right and left lungs
Windpipe	b	h.	A short tube stiffened by rings of cartilage

2. Which figure shows a person inhaling? Which figure shows a person exhaling?

    a. exhaling

    b. inhaling

Conceptual Integrated Science — Third Edition

Chapter 20: Human Biology II—Care and Maintenance
Digestion

1. Why do we have to digest our food?

   Food has to be broken down into smaller organic molecules

   that can be absorbed and used by the body.

2. Label the parts of the digestive system in the diagram using the following terms:

Stomach

Liver

Pancreas

Small intestine

Esophagus

Large intestine

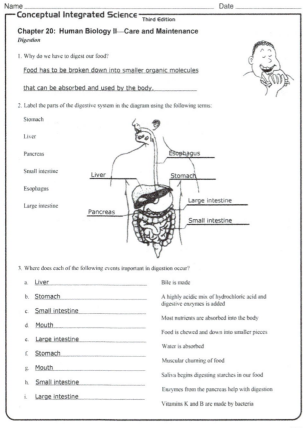

3. Where does each of the following events important in digestion occur?

a.	Liver	Bile is made
b.	Stomach	A highly acidic mix of hydrochloric acid and digestive enzymes is added
c.	Small intestine	Most nutrients are absorbed into the body
d.	Mouth	Food is chewed and down into smaller pieces
e.	Large intestine	Water is absorbed
f.	Stomach	Muscular churning of food
g.	Mouth	Saliva begins digesting starches in our food
h.	Small intestine	Enzymes from the pancreas help with digestion
i.	Large intestine	Vitamins K and B are made by bacteria

Conceptual Integrated Science — Third Edition

Chapter 20: Human Biology II—Care and Maintenance
Excretory System

1. Label the parts of the excretory system in the diagram using the following terms:

Urethra

Bladder

Ureter

Kidney

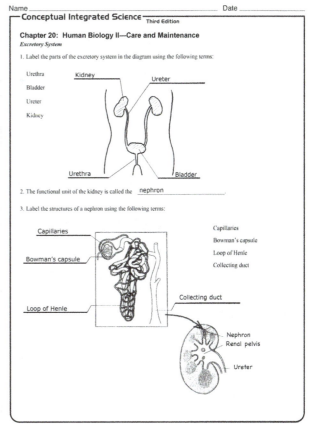

2. The functional unit of the kidney is called the ___nephron___.

3. Label the structures of a nephron using the following terms:

Capillaries

Bowman's capsule

Loop of Henle

Collecting duct

Conceptual Integrated Science — Third Edition

Chapter 20: Human Biology II—Care and Maintenance
Excretory System—continued

4. Fluid leaves the ___capillaries___ of the circulatory system and enters the

   part of the nephron called ___Bowman's capsule___.

5. The two parts of the nephron where water is reabsorbed by the body are the

   ___loop of Henle___ and the ___collecting duct___.

> The attention a community gives to the maintenance of its infrastructure is a measure of the value of that community. Likewise with the care and maintenance of your body.

Conceptual Integrated Science — Third Edition

Chapter 20: Human Biology II—Care and Maintenance
Immune System

1. Describe the role of each part of the immune system.

   a. Skin keeps pathogens from getting into the body.

   b. Mucus A layer of mucus covers all our mucus membranes. Mucus helps trap pathogens.

   c. Enzymes in tears and milk kill bacteria.

   d. Inflammatory response In the inflammatory response, swelling helps isolate wounds. In addition, innate immune cells attack pathogens that have entered the body.

   e. Antibodies Antibodies bind to antigens on pathogens. This prevents the pathogens from functioning properly or causes antigens to clump together, making them easy targets for other immune cells.

   f. Clones Both B and T cells produce clones, many copies of themselves, for strengthening the immune response.

   g. B cells B cells attack pathogens in bodily fluids (blood or lymph).

   h. T cells T cells attack pathogens that are inside the body's cells.

   i. Memory cells Memory cells remain in the body for a long time after an infection. If the same pathogen attacks the body again, the memory cells start an immediate, aggressive attack.

2. Explain how a vaccine works. Most vaccines contain either dead pathogens or weak versions of a pathogen, or they use only a part of a pathogen—maybe part of a virus's protein coat or part of a bacterium's cell wall. The acquired immune system reacts to antigens in the vaccine just as it would react to the real pathogen. It makes antibodies and —most importantly—memory cells. If the real pathogen ever shows up, the acquired immune system is ready to launch an immediate, aggressive attack.

Conceptual Integrated Science *Third Edition*

Chapter 21: Ecology
Populations, Communities, and Ecosystems

1. Define the following terms.

Ecology   <u>the study of how organisms interact with their environments</u>

Population   <u>a group of individuals of a single species that lives in a specific area</u>

Community   <u>all the organisms that live in a specific area</u>

Ecosystem   <u>all the organisms that live in a specific area and all the abiotic features of</u>
<u>their environment</u>

2. In the space below, draw three (or more) populations of organisms in their habitat. Then, add labels to your drawing to identify:

a. each of the three populations

b. at least one way in which two of the populations interact with each other

c. at least one way in which one population interacts with the abiotic features of its environment

(For example, you might draw plants, butterflies, and birds in your backyard. You could then describe an interaction between the butterflies and plants in which butterflies feed on nectar from the plant's flowers. Finally, you could explain how the birds drink water from a fountain in the yard. Come up with your own example!)

Many answers are possible. This is just one example.

Robins eat worms.
(an interaction between two species)

| Worms | → | Robins | | Hummingbirds |

Live in the soil.
(an interaction with abiotic features of the environment)

If this, then that.
That's scientific thinking.

Conceptual Integrated Science *Third Edition*

Chapter 21: Ecology
Exponential Growth and Logistic Growth

1. Which of the following graphs shows exponential growth and which shows logistic growth?

exponential growth   logistic growth

Circle the correct answers:

2. The carrying capacity in the graph on the right is (10) ((100)) (more than 100) individuals.

Appendix D will help you with Questions 3 and 4.

3. Exponential growth is nicely illustrated with the children's story of a rapidly growing beanstalk that doubles in height each day.

Suppose one day after breaking ground the stalk is 1 centimeter high.

If growth is continual, at the end of the second day it will be (1) ((2)) (4) cm high.

At the end of the third day it will be (1) (2) ((4)) cm high.

Doubling each day results in exponential growth so that on the 36th day it reaches the Moon! Working backward, the height of the beanstalk on the 35th day was ((one-half)) (one-quarter) (one-third) the distance from Earth to the Moon.

And on the 34th day the beanstalk was (one-half) ((one-quarter)) (one-third) the distance from Earth to the Moon.

4. Then there is the story of a lily pond with a single leaf. Each day the number of leaves doubles; on the 30th day the pond is completely full.

On what day was the pond half covered? (15 days) (28 days) ((29 days))

On what day was it one-quarter covered? (15 days) ((28 days)) (29 days)

Conceptual Integrated Science *Third Edition*

Chapter 21: Ecology
Species Interactions

Circle the correct answers:
Look at the food chain below.

coyotes

lizards

insects   rodents   deer

plants

a. The producers in this community are (insects) ((plants)) (rodents).

b. The primary consumers in this community are
(plants, insects, and rodents) ((insects, rodents, and deers)) (insects, lizards, and coyotes).

c. The secondary consumers in this community are
(plants, insects, and rodents) (insects, rodents, and deers) ((lizards and coyotes)).

d. The tertiary consumers in this community are (deer) ((coyotes)) (insects).

e. The top predators in this community are (deer) ((coyotes)) (insects).

┌─ **Conceptual Integrated Science** ─ Third Edition

Chapter 21: Ecology
Species Interactions—continued

3. Briefly describe a species' niche.

   It is the total set of biotic and abiotic resources a species uses within a community.

   _____

Fill in the blanks:

4. In a symbiosis, individuals of two species live in close association with one another.

   a. The three types of symbiosis are   parasitism   ,

      commensalism   ,   and   mutualism   .

Circle the correct answers:

   b. ((Parasitism)) (Commensalism) (Mutualism)   is good for one member of the interaction

      and bad for the other.

   c. (Parasitism) ((Commensalism)) (Mutualism)   is good for one species in the interaction

      and has no effect on the other.

   d. (Parasitism) (Commensalism) ((Mutualism))   is a relationship that benefits both species

      involved.

┌─ **Conceptual Integrated Science** ─ Third Edition

Chapter 21: Ecology
Biomes and Aquatic Habitats

1. Match each of the following features with the appropriate biome:

a. Tropical grassland	f _____	Tropical forest
b. A habitat that receives very little precipitation, can be cold or hot	h _____	Temperate forest
	e _____	Coniferous forest
c. Permafrost is found in this biome	c _____	Tundra
d. Mild, rainy winters, and hot summers with drought and fire	a _____	Savanna
e. The trees in this biome have needlelike leaves that can survive cold winters	g _____	Temperate grassland
	b _____	Desert
f. More species are found in this biome than in all other biomes combined	d _____	Chaparral
g. A grassland found in areas with four distinct seasons		
h. In this biome, trees drop their leaves in the autumn		

2. What kinds of adaptations do you expect to see in a freshwater organism that lives in a river or stream? Why?

   Species that live in the flowing waters of rivers and streams usually have adaptations

   that keep them from being washed away. Many have hooks or suckers for attaching

   to rocks. Others are strong swimmers.

3. a. What is an estuary?

   Estuaries are habitats where freshwater rivers flow into oceans.

   b. What adaptations are found in plants that live in estuaries?

   Plants that live in estuaries have adaptations that allow them to deal with

   changing salt levels.

┌─ **Conceptual Integrated Science** ─ Third Edition

Chapter 21: Ecology
Biomes and Aquatic Habitats—continued

4. a. What is an intertidal habitat?

   Intertidal habitats are oceanic habitats found close to shore.

   b. What feature of intertidal habitats makes them challenging environments to live in?

   As the tide moves in and out, intertidal habitats alternate between being under

   water and exposed to air. Organisms that live in intertidal habitats also have to

   deal with changes in temperature and waves.

   c. What are some adaptations found in organisms that live in intertidal habitats?

   Many intertidal species have thick shells or hide in crevices to keep from drying

   out. In addition, all species can attach firmly to rocks or other surfaces so that

   they don't get washed up onto the beach.

Chemistry is applied physics. Biology is applied physics *and* chemistry. Earth science applies *all*!

Name_____ Date _____

┌─ Conceptual Integrated Science ── Third Edition

Chapter 21: Ecology
Carbon Cycle and Nitrogen Cycle

1. What is a biogeochemical cycle? Why does the term include both "bio" and "geo"?

   A biogeochemical cycle describes how substances such as water or carbon, nitrogen,

   and other elements move around Earth, going back and forth between "bio" (living

   organisms) and "geo" (Earth's atmosphere, crust, and waters).

2. Draw one possible path of a carbon atom from the atmosphere, to a plant, to an animal, and back to the atmosphere. In your drawing, indicate where "photosynthesis" occurs and where "cellular respiration" occurs.

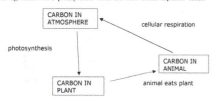

3. How does nitrogen move from the abiotic world to the biotic world?

   Nitrogen-fixing bacteria convert nitrogen gas to ammonium and then nitrifying bacteria

   convert it to nitrates, a form plants can easily use.

4. How does nitrogen move from the biotic world to the abiotic world?

   Denitrifying bacteria convert nitrogen-containing compounds back to nitrogen gas.

214

Name_____ Date _____

┌─ Conceptual Integrated Science ── Third Edition

Chapter 21: Ecology
Ecological Succession

1. What is ecological succession?

   Ecological succession describes how a community of organisms in a habitat changes

   over time.

2. What are some differences between primary succession and secondary succession?

   Primary succession occurs when bare land (with no soil) becomes inhabited by waves of

   organisms. In secondary succession, land with soil is colonized. Secondary succession

   usually occurs much more quickly than primary succession.

3. What is a pioneer species?

   Pioneer species are the first species to colonize bare land. They are able to survive with

   new nutrients and little existing organic material. They are often exposed to variable

   sunlight and temperatures.

4. What is a climax community?

   A climax community is the relatively stable final community (set of living organisms) that

   is found at the end of the process of ecological succession.

5. Hoes does biodiversity change during ecological succession?

   Biodiversity typically increases over time during ecological succession.

6. Explain the intermediate disturbance hypothesis.

   The intermediate disturbance hypothesis suggests that small, regular disturbances in a

   habitat can actually increase biodiversity in that habitat. This is because disturbed

   patches of habitat can sometimes be used by different organisms (that are not normally

   part of the climax community) as these patches go through different stages of

   recovery.

215

Name_____ Date _____

┌─ Conceptual Integrated Science ── Third Edition

Chapter 22: Plate Tecton: The Earth System
How Many Layers?

The three-layer model

When you were younger, you may have learned that Earth has *three* layers: crust, mantle, and core. This simple model is based on Earth's *composition*, or chemical makeup. But Earth scientists (and science students like you) know that there is another way to look at our planet. Scientists usually divide Earth into five structural layers on the basis of physical properties: *lithosphere, asthenosphere, mesosphere, outer core,* and *inner core.*

1. a. Label the three layers:

   Crust

   Mantle

   Core

   Crust

   Mantle

   Core

Fill in the blanks:

   b. In this three-layer model, Earth's surface layer is called the   crust   .

   It consists mostly of low-density rocks such as   granite   and

   basalt   .

   Earth's middle layer is called the   mantle   . It consists of

   rocks   that contain dark and dense elements such as magnesium.

   Earth's innermost layer is mostly made of   iron   .

217

Name_____ Date _____

┌─ Conceptual Integrated Science ── Third Edition

Chapter 22: Plate Tecton: The Earth System
How Many Layers?—continued

The five-layer model

1. a. Label the five layers:

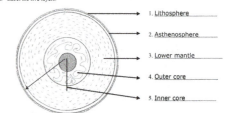

   1. Lithosphere

   2. Asthenosphere

   3. Lower mantle

   4. Outer core

   5. Inner core

   b. Describe the physical properties of each layer.

   Lithosphere: cool, rigid rock

   Asthenosphere: warmer, plastic (slowly flowing) rock

   Lower mantle: strong and rigid rock

   Outer core: hot, liquid metal

   Inner core: extremely hot solid metal under enormous pressure

Circle True or False:

2. a. The asthenosphere is a soft layer of the mantle on which pieces of lithosphere move.   (True) (False)

   b. The lithosphere is made of two parts—the crust and the upper mantle.   (True) (False)

   c. Even though the lithosphere is made of different types of rock, it behaves as a single unit.   (True) (False)

   d. The lithosphere is a rigid layer of brittle rock.   (True) (False)

In the drawing above do the following:

3. a. With a red crayon or colored pencil, color the hottest layer of Earth.   (Layer 5)

   b. With a blue crayon or colored pencil, color the coolest part of Earth.   (Layer 1)

   c. Draw an arrow to show the direction of net heat flow within Earth.

218

320

Conceptual Integrated Science — Third Edition

Chapter 22: Plate Tecton: The Earth System
Get the Picture—Plate Boundaries

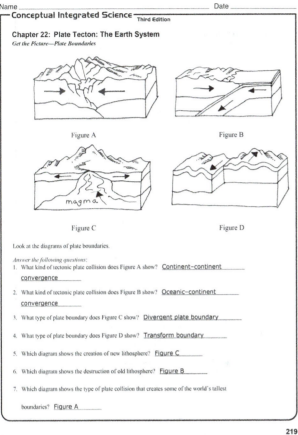

Figure A   Figure B

Figure C   Figure D

Look at the diagrams of plate boundaries.

Answer the following questions:

1. What kind of tectonic plate collision does Figure A show? Continent–continent convergence

2. What kind of tectonic plate collision does Figure B show? Oceanic–continent convergence

3. What type of plate boundary does Figure C show? Divergent plate boundary

4. What type of plate boundary does Figure D show? Transform boundary

5. Which diagram shows the creation of new lithosphere? Figure C

6. Which diagram shows the destruction of old lithosphere? Figure B

7. Which diagram shows the type of plate collision that creates some of the world's tallest boundaries? Figure A

Conceptual Integrated Science — Third Edition

Chapter 22: Plate Tecton: The Earth System
Get the Picture—Plate Boundaries—continued

8. What geological events occur most often near plate boundaries?

Earthquakes, mountain building, volcanic eruptions

9. Why are plate boundaries active geological regions?

Plates collide and rub against one another and these motions put great stresses on the lithosphere. Also, magma wells up at (divergent) plate boundaries, which pushes the old lithosphere aside and builds volcanic mountains.

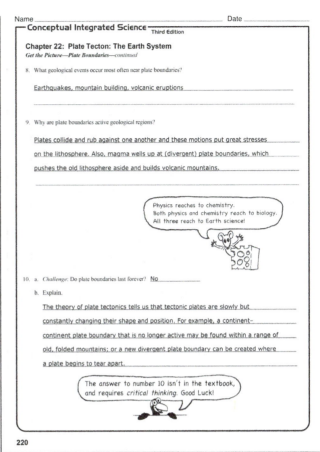

Physics reaches to chemistry.
Both physics and chemistry reach to biology.
All three reach to Earth science!

10. a. *Challenge*: Do plate boundaries last forever? No

b. Explain.

The theory of plate tectonics tells us that tectonic plates are slowly but constantly changing their shape and position. For example, a continent–continent plate boundary that is no longer active may be found within a range of old, folded mountains; or a new divergent plate boundary can be created where a plate begins to tear apart.

The answer to number 10 isn't in the textbook, and requires *critical thinking*. Good Luck!

Conceptual Integrated Science — Third Edition

Chapter 22: Plate Tecton: The Earth System
Get the Picture—Seafloor Spreading

1. The diagram shows what happens during seafloor spreading. But it isn't very useful unless you can label all the parts. Match the numbers (1 to 5) to the terms below.

4 convection
5 magma rising
3 mid-ocean ridge
1 older crust
2 younger crust

2. During seafloor spreading, magma rises to fill a gap between diverging tectonic plates.
After magma erupts it is called *lava*.
Draw in the lava that would erupt at this spreading center.

3. What happens to lava after it cools on the seafloor?

It solidifies to become rock—mainly basalt.

4. Why is the process in the diagram called *seafloor spreading*?

The seafloor literally spreads apart: As new lithosphere is created and pushes the older lithosphere away from a spreading center, the seafloor becomes wider.

_____

Conceptual Integrated Science — Third Edition

Chapter 22: Plate Tecton: The Earth System
Play Tectonics

Use the clues below to fill in the words of the crossword puzzle in the numbered blanks on the facing page.

Across

1. The fundamental theory of Earth Science

5. The mode of heat transfer that makes Earth's mantle churn

7. The global system of undersea mountains, called the *mid-ocean* ridge

8. Plate boundaries where tectonic plates slide past one another

9. Inside ocean trenches, the temperature is very cold .

10. German scientist who advanced the notion of continental drift

12. The process by which gravity contributes to plate motion

14. The recycling of lithosphere at an ocean trench

Down

2. Earth's rigid layer, which is broken up into tectonic plates

3. What gets wider when new seafloor is created?

4. Earth's plastic layer, which tectonic plates slide over

6. What escapes from Earth's interior and drives the motion of tectonic plates?

11. The branch of Earth Science concerned with Earth's interior

13. Another name for chunks of lithosphere

15. Old lithosphere sinks at ocean trenches because it is cooler and therefore more dense than the young lithosphere forming near spreading centers.

16. A deep crack in the ocean floor where old lithosphere is subducted

Conceptual Integrated Science — Third Edition

Chapter 22: Plate Tecton: The Earth System
Play Tectonics—continued

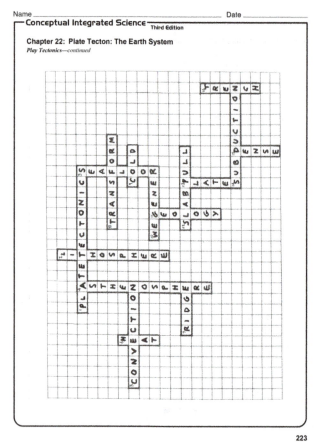

Conceptual Integrated Science — Third Edition

Chapter 22: Plate Tecton: The Earth System
Plate Tectonics Vocabulary Review

Fill in the blanks with the correct word:

1. The asthenosphere is the layer of the Earth located just below the **lithosphere** (core, lithosphere, mantle).

2. The layer of Earth that makes up most of Earth's mass is the **mantle** (mantle, crust, lithosphere).

3. The layer of Earth that is made up of tectonic plates is the **lithosphere** (asthenosphere, crust, lithosphere).

4. Tectonic plates move because they are carried on top of the slowly flowing **asthenosphere** (asthenosphere, crust, lithosphere).

5. Earth's core formed when iron sank to Earth's center when the planet was young and molten. Iron sank because it is **dense** (magma, dense, subducting).

6. New lithosphere is created in the process of **seafloor spreading** (continental drift, seafloor spreading, convection).

7. Old lithosphere is recycled at a(n) **ocean trench** (transform boundary, ocean trench, tectonic plate).

8. If you explore the ocean floor in a deep-sea submarine, you might see the **mid-ocean ridge** (mid-ocean ridge, asthenosphere, mantle).

9. Mountains are sometimes made where tectonic plates crash into one another at a **convergent boundary** (divergent boundary, transform boundary, convergent boundary).

10. The youngest, warmest, and least dense part of Earth's crust exists at a **divergent boundary** (divergent boundary, transform boundary, convergent boundary).

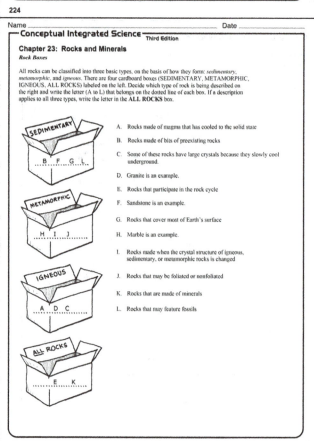

> lithosphere crust
> tectonic plates
> mantle

> Fellow scientists once ignored my hypothesis of _continental drift_
> Later they based the theory of _plate tectonics_ on my work.
> I was ahead of my time!

Conceptual Integrated Science — Third Edition

Chapter 23: Rocks and Minerals
Mineral Detective

Use the clues given in Sections 23.2 and 23.3 of your textbook to identify each mineral.

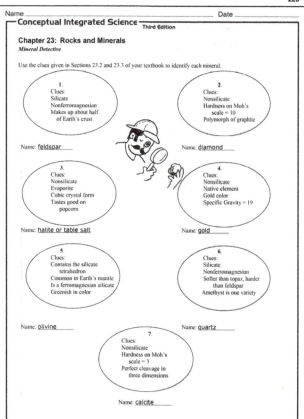

1.
Clues:
Silicate
Nonferromagnesian
Makes up about half of Earth's crust

Name: feldspar

2.
Clues:
Nonsilicate
Hardness on Moh's scale = 10
Polymorph of graphite

Name: diamond

3.
Clues:
Nonsilicate
Evaporite
Cubic crystal form
Tastes good on popcorn

Name: halite or table salt

4.
Clues:
Nonsilicate
Native element
Gold color
Specific Gravity = 19

Name: gold

5.
Clues:
Contains the silicate tetrahedron
Common in Earth's mantle
Is a ferromagnesian silicate
Greenish in color

Name: olivine

6.
Clues:
Silicate
Nonferromagnesian
Softer than topaz, harder than feldspar
Amethyst is one variety

Name: quartz

7.
Clues:
Nonsilicate
Hardness on Moh's scale = 3
Perfect cleavage in three dimensions

Name: calcite

Conceptual Integrated Science — Third Edition

Chapter 23: Rocks and Minerals
Rock Boxes

All rocks can be classified into three basic types, on the basis of how they form: *sedimentary, metamorphic,* and *igneous.* There are four cardboard boxes (SEDIMENTARY, METAMORPHIC, IGNEOUS, ALL ROCKS) labeled on the left. Decide which type of rock is being described on the right and write the letter (A to L) that belongs on the dotted line of each box. If a description applies to all three types, write the letter in the **ALL ROCKS** box.

SEDIMENTARY: B F G L

METAMORPHIC: H I J

IGNEOUS: A D C

ALL ROCKS: E K

A. Rocks made of magma that has cooled to the solid state

B. Rocks made of bits of preexisting rocks

C. Some of these rocks have large crystals because they slowly cool underground.

D. Granite is an example.

E. Rocks that participate in the rock cycle

F. Sandstone is an example.

G. Rocks that cover most of Earth's surface

H. Marble is an example.

I. Rocks made when the crystal structure of igneous, sedimentary, or metamorphic rocks is changed

J. Rocks that may be foliated or nonfoliated

K. Rocks that are made of minerals

L. Rocks that may feature fossils

Tinkering is the process of "nosing" your way toward a solution to something you don't quite know how to do — a combination of discovery and play.

Conceptual Integrated Science
Third Edition

Chapter 23: Rocks and Minerals
Are You a Rock Hound?

Are rocks your hobby? If you know a lot about rocks, do this crossword puzzle on the facing page and see if you qualify as an official "rock hound."

Use the clues below to fill in the words of the crossword puzzle in the numbered blanks.

Across

1. What determines a mineral's crystal form?
5. Over half of Earth's crust is made of this mineral.
6. What class of minerals includes quartz, feldspar, and others that contain silicon and oxygen?
8. Most minerals form when this cools to the solid state.
9. A rock is a solid <u>mixture</u> of minerals.
11. Rocks that are formed when preexisting rocks recrystallize due to temperature or pressure changes
12. Intrusive igneous rocks, such as granite, form here.

Down

2. Minerals of the same kind have the same crystal structure and <u>chemical</u> composition.
3. Minerals that exhibit this property contain planes of atoms that are weakly bonded to one another.
4. This mineral, the second most common in Earth's crust, is clear when pure but can be yellow, purple, white, pink, or brown when it contains impurities.
7. What chemical element makes up almost half of Earth's crust, yet most people think it is a gas found in air?
10. Rocks formed by magma that cools and solidifies, making mineral crystals that stick together
13. A mineral deposit that is large and pure enough to be mined for a profit
14. A tunnel, pit, or strip of bulldozed earth from which minerals are removed

---

Conceptual Integrated Science
Third Edition

Chapter 23: Rocks and Minerals
Are You a Rock Hound?—continued

Do geologists who wonder have rocks in their heads?

The five E's of education: Engage, Explore, Explain, Extend, and Evaluate.

Conceptual Integrated Science — Third Edition

Chapter 24: Earth's Surface—Land and Water
Fault Facts

OUCH!

Imagine you are tiny like Perky the mouse. If you fell into a fault during an earthquake, what would it look like? How would the slipping blocks of rock move?
Answer: It depends on the fault. There are three basic kinds of faults. Rock moves differently along each fault type. Explore fault types here and in Section Integrated Science (IS) 24A of your textbook.

Normal Fault
Faults are classified by how the two rocky blocks on either side of the fault move relative to one another. In a normal fault, rock on one side of the fault drops *down*, relative to the other side. Investigate the diagram now.

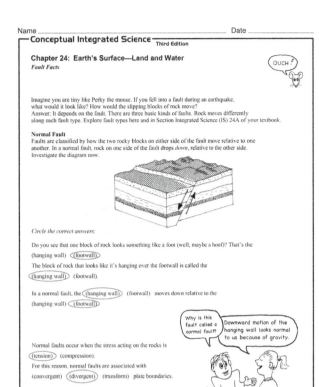

Circle the correct answers:

Do you see that one block of rock looks something like a foot (well, maybe a hoof)? That's the
(hanging wall) (footwall)
The block of rock that looks like it's hanging over the footwall is called the
(hanging wall) (footwall).

In a normal fault, the (hanging wall) (footwall) moves down relative to the
(hanging wall) (footwall)

Why is this fault called a *normal fault*?

Downward motion of the hanging wall looks normal to us because of gravity.

Normal faults occur when the stress acting on the rocks is
(tension) (compression).
For this reason, normal faults are associated with
(convergent) (divergent) (transform) plate boundaries.

Conceptual Integrated Science — Third Edition

Chapter 24: Earth's Surface—Land and Water
Fault Facts—continued

Reverse Fault
Along a reverse fault, one block of rock is pushed *up* relative to the other side. This is the reverse of what gravity would do, correct?

Circle the correct answers:
When movement along a fault is the reverse of what normal gravity would produce, we call the
fault a (normal) (reverse) fault.
Reverse faults occur when the stress acting on the rocks is
(tension) (compression).
For this reason, normal faults are associated with
(convergent) (divergent) (transform) plate boundaries.

Strike-slip Fault

Circle the correct answer:
Blocks of rock move up or down along normal and reverse faults. But strike-slip faults are different. The gigantic blocks of rock on either side of a strike-slip fault scrape beside one another. Motion is horizontal, not vertical. Strike-slip boundaries occur at
(convergent) (divergent) (transform) plate boundaries.

How Faults Affect the Strength of Earthquakes
Earthquakes generally occur at faults. The strength of an earthquake depends on the kind of faults involved. Fill in the blanks below showing how fault type relates to earthquake strength.

Plate boundary	Major fault type	Earthquake strength
transform	strike-skip fault	moderate
convergent	reverse	strong
divergent	normal fault	weak

Conceptual Integrated Science — Third Edition

Chapter 24: Earth's Surface—Land and Water
The Story of Old, Folded Mountains—The Appalachians

Folded mountains form when blocks of rock are squeezed together and push upward like wrinkled cloth. Usually folded mountains occur at convergent plate boundaries, where tectonic plates collide. The Appalachian Mountains, however, lie in eastern North America in the middle of the North American plate. How can this be? *Find out how the Appalachian Mountains developed. Then circle the answers on the next page to tell their story.*

500 million years ago

North America   Europe

Africa

1. The landmasses that would become North America and Africa were moving toward one another about 500 million years ago.

B A M !

400 million years ago

Appalachian Mountains

2. About 400 million years ago—BAM! The tectonic plates that North America and Africa were riding on collided. The huge collision caused the crust to fold upward, creating the Appalachian Mountains.

65 million years ago

North America   Atlantic Ocean   Africa

3. About two million years later, North America and Africa began to split. As the Atlantic Ocean grew wider, the Appalachians were shifted away from the plate boundary. By 65 million years ago, the Appalachians had moved to a plate interior. No longer were the Appalachians near a plate boundary.

Conceptual Integrated Science — Third Edition

Chapter 24: Earth's Surface—Land and Water
The Story of Old, Folded Mountains—The Appalachians—continued

PLATE TECTONICS =SIGH!

Circle the correct answers:

1. The Appalachian Mountains are (old) (young) mountains.
   You can tell this by their low, rounded shapes.

2. The Appalachians are
   (folded) (fault-block) (volcanic) mountains.

3. The forces that produced the Appalachians were due to
   (tension) (compression)

4. About 400 million years ago the Appalachians formed when the landmasses that would be
   North America and Africa (collided) (pulled apart).

5. So, the Appalachian Mountains, like most other folded mountains, developed at a
   (divergent) (convergent) plate boundary.

6. Then, about 200 million years ago, North America and Africa began to break apart.
   The plate boundary that formed between them was (divergent) (convergent).

7. In summary, the Appalachians formed at the boundary of two
   (converging) (diverging) plates. Later, a divergent boundary formed in the Appalachian mountain range. This boundary split the mountains. Some of the mountains drifted along with Africa and some drifted along with North America. New oceanic lithosphere—the Atlantic Ocean basin—was created where the plates were splitting apart. As the seafloor got wider over time, the Appalachians were shifted toward the
   (edge) (interior) of the North American Plate. And that's the story of how the Appalachian Mountains came to be located in the middle of a
   (tectonic plate) (mid-ocean ridge).

Are plates still moving?

Experimental evidence is the test of truth in science.

Conceptual Integrated Science
Third Edition

Chapter 24: Earth's Surface—Land and Water
Where the Action Is

1. Investigate the map of the world that shows the main areas where folded mountains exist.

KEY
⫽ Main areas of folded mountains

Are folded mountains randomly placed around the globe?

No, the folded mountains occur in swaths.

2. Now investigate this map of the world that shows the main areas of volcanic activity and where earthquakes occur.

KEY
▨ Region where most earthquakes happen
• Volcano

Conceptual Integrated Science
Third Edition

Chapter 24: Earth's Surface—Land and Water
Where the Action Is—continued

  a. Do earthquakes and volcanic activity generally occur in the same places?   Yes

  b. If so, why would this be true?

    Both earthquakes and volcanism are associated with plate tectonic boundaries.

3. Compare the two maps on the facing page.

  a. Can you see the pattern?   Yes

  b. Briefly give an explanation for the pattern.

    Folded mountains are associated with earthquakes and the presence of

    volcanoes.

4. Now investigate the map that shows the main tectonic plates. Relate the location of folded mountains, earthquakes, and volcanic activity to tectonic plate boundaries.

KEY
╱ Plate boundary
→ Direction that plate is moving

What pattern do you see?

A comparison of all the maps shows that folded mountains, earthquakes, and

volcanoes exist along plate boundaries.

Conceptual Integrated Science
Third Edition

Chapter 24: Earth's Surface—Land and Water
Where the Action Is—continued

5. Explain why folded mountains, earthquakes, and volcanic activity are most common near tectonic plate boundaries.

    Earthquakes occur where rock sections grind together, stick, then slip, which

    occurs at plate boundaries. Similarly, volcanism is usually caused by the upwelling

    of magma that occurs at divergent plate boundaries and at the subduction zones,

    where sinking plates produce melting of rock. Folded mountains are associated

    with convergent plate boundaries.

6. Why was there little understanding of how Earth's surface features developed before the 1960s? [Hint: When was the theory of Plate Tectonics developed?]

    Prior to the 1960s there was no theory of Plate Tectonics; hence, surface features

    that arise from the motion of Earth's plates could not be explained.

Most of geology's action is in slow motion.

Conceptual Integrated Science — Third Edition

Chapter 24: Earth's Surface—Land and Water
What Do You Know About Water Flow?

Look at the diagram of the water cycle. Next to each reservoir, write how long water stays in that reservoir.

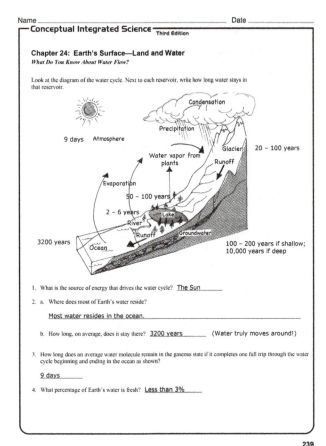

1. What is the source of energy that drives the water cycle? __The Sun__

2. a. Where does most of Earth's water reside?

   __Most water resides in the ocean.__

   b. How long, on average, does it stay there? __3200 years__   (Water truly moves around!)

3. How long does an average water molecule remain in the gaseous state if it completes one full trip through the water cycle beginning and ending in the ocean as shown?

   __9 days__

4. What percentage of Earth's water is fresh? __Less than 3%__

Conceptual Integrated Science — Third Edition

Chapter 24: Earth's Surface—Land and Water
What Do You Know About Water Flow?—continued

5. a. In what form is most of Earth's fresh water?

   __Frozen as snow and glacial ice__

   b. In what part of the world is it located?

   __In the polar regions, where it is the coldest__

6. When precipitation doesn't evaporate or soak into the ground, it becomes runoff.

   Where does runoff eventually go? __Ocean__

7. What percentage of Earth's water takes part in the water cycle? __100%__

8. a. Does the water in your body take part in the water cycle? __Yes__

   b. Explain.

   __Water stored in organisms makes up a tiny portion of the hydrosphere—about__
   __0.00004% of it. A human being comprises about 70% water. When one dies, the water__
   __in the body returns to Earth and flows through its reservoirs. Further, water flows in and__
   __out of your body by drinking, urinating, perspiring, etc. All this water flows through the__
   __oceans, atmosphere, ground, and other reservoirs over geologic time.__

9. On a typical day in the United States, about 4 trillion gallons of precipitation falls. What happens to this water once it falls?

   __Most evaporates; The liquid water that doesn't evaporate percolates underground__
   __to become groundwater or flows toward the ocean as runoff.__

Bottled water costs as much and often more than fruit juice?

Conceptual Integrated Science — Third Edition

Chapter 25: Surface Processes
Weathering Earth's Crust

Weathering is the process by which rocks in Earth's crust are broken down into smaller pieces. These pieces vary in size from boulders to pebbles to the tiny particles that make up soil.

Complete the concept map by filling in the blanks:

```
                    Weathering
            is the breakdown of rock.
            It happens in two ways:

   Mechanical weathering    and    Chemical weathering
is the breakdown of rock by physical      is the breakdown of rock by chemical
processes.                                processes.
Examples include                         Examples include

     ice wedging                          dissolving by water

     wind abrasion                        corrosion by acid rain

  intrusion of plant roots                     rusting
```

Conceptual Integrated Science — Third Edition

Chapter 25: Surface Processes
Weathering Earth's Crust—continued

1. Give two examples of weathering you have observed at home or at school over the past week. Describe how the weathering occurred for each example.

   First example: __Answers will vary. Potholes in pavement: These can form by water__
   __seeping below pavement then freezing and expanding and pushing upward and out-__
   __ward on the pavement.__

   Second example: __Rounded or cracked rocks: These form by mechanisms that include__
   __ice wedging, abrasion, and reactions with water.__

2. a. Rocks expand and contract due to changes in temperature. Over time, temperature variations crack rocks and break them up. Where else might you see weathering due to temperature changes? [Hint: You walk on it everyday.]

   __Streets and sidewalks show cracks because of ice wedging and volume changes__
   __that arise from thermal expansion in warm weather.__

   b. Is this kind of weathering mechanical or chemical? __Mechanical weathering__

3. Give an example of *mechanical* weathering caused by water. __Answers will vary.__

   __Example: Breakup of a cliff along the seashore as ocean waves pound it during a__
   __storm.__

4. Give an example of *chemical* weathering caused by water.

   __The corrosion of a marble statue due to acid rain, for example.__

5. What is the final product of weathering? __Soil__

6. a. What is soil? __It is a mixture of fine, weathered rock particles and air, water, and__
   __organic matter such as dead animals and plants and animal excrement.__

   b. Why does it take so long to form?

   __Highly weathered rock in it takes a long time to decompose.__

7. How is the process of weathering involved in the growing of crops?

   __Crops grow in soil, and a component of soil is weathered rock. Rock indirectly__
   __nourishes crops because minerals from weathered rock can provide nutrients.__

Conceptual Integrated Science *Third Edition*

Chapter 25: Surface Processes
The Speed of Water Affects Sedimentary Rock Formation

Some rocks form when rock fragments are squeezed together under pressure—we say the rock is *compacted*. Three common sedimentary rocks formed this way are shown in the table. Note that shale is made of tiny particles, sandstone is made of particles the size of sand grains, and conglomerate is made of much bigger particles.

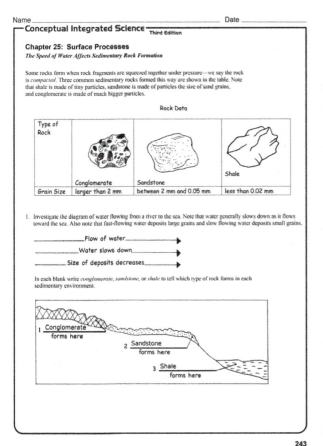

Rock Data

Type of Rock	Conglomerate	Sandstone	Shale
Grain Size	larger than 2 mm	between 2 mm and 0.05 mm	less than 0.02 mm

1. Investigate the diagram of water flowing from a river to the sea. Note that water generally slows down as it flows toward the sea. Also note that fast-flowing water deposits large grains and slow flowing water deposits small grains.

_____ Flow of water _____ ➤
_____ Water slows down _____ ➤
_____ Size of deposits decreases _____ ➤

In each blank write *conglomerate*, *sandstone*, or *shale* to tell which type of rock forms in each sedimentary environment.

1 __Conglomerate__ forms here
2 __Sandstone__ forms here
3 __Shale__ forms here

Conceptual Integrated Science *Third Edition*

Chapter 25: Surface Processes
The Speed of Water Affects Sedimentary Rock Formation—continued

Fill in the blanks by choosing the correct answer.

2. The faster water flows, the more energy it has and the bigger the particles it can carry.

Water generally moves

__slower__ (faster, slower) as it flows from steep mountains toward flatter land

near the sea. As the water slows down, the particles it first deposits are

__bigger__ (bigger, smaller).

A sedimentary rock made of large particles, such as stones and pebbles, is

__conglomerate__ (conglomerate, sandstone, shale). So this rock forms along river

beds where rushing water has lost enough of its speed to begin dropping its larger sediments.

Water tends to slow down near the mouth of a river, when the river meets the sea. So water

deposits medium-sized grains, such as sand, near a river's mouth. So near the mouth of a

river we find

__sandstone__ (conglomerate, sandstone, shale).

But the smallest particles, such as clay, aren't deposited until they reach the sea. In the sea,

where water has little energy of motion, small particles are finally deposited. Fine sediments

are composed of rocks such as

__shale__ (conglomerate, sandstone, shale) that develop along the still sea

bottom over geologic time.

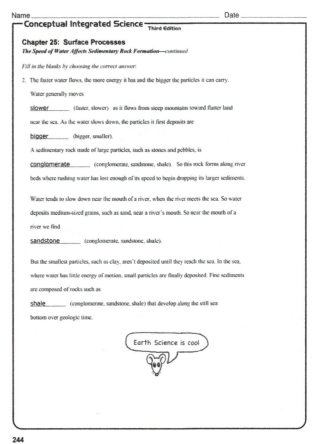

Earth Science is cool

Conceptual Integrated Science *Third Edition*

Chapter 25: Surface Processes
Landforms Created by Erosion

1. a. Label each landform by filling in the blanks.

Delta
Floodplain
Headwaters
Levee
Meander
Mouth
Tributary
V-shaped valley

1 __V-shaped valley__
6 __Headwaters__
7 __Tributary__
2 __Meander__
3 __Floodplain__
4 __Levee__
8 __Mouth__
5 __Delta__

b. Describe each landform.

Delta __A delta is a mass of sediment deposited where a river slows as it enters a large body of water such as a lake or the ocean. When a stream reaches the coastline, the gradient is usually low and stream velocity slow. Only the small sediments remain, so deltas are deposits of finer sediments including gravels and sand.__

Floodplain __This is the flat area along a stream covered by water only during a flood. When a stream floods, the water overtopping the its banks slows as it flows away from the stream channel. Sediments from the stream drop out of the slower water and build up a flat floodplain. Floodplains are especially fertile for farming because of the fresh soil deposited there during recurring floods.__

Conceptual Integrated Science *Third Edition*

Chapter 25: Surface Processes
Landforms Created by Erosion—continued

Headwaters __A stream's headwaters are the source of its waters.__

Levee __A natural levee is a deposit of sand or mud that builds up in a floodplain beside a stream. Levees are built up from sediments deposited during repeated episodes of flooding.__

Meander __A meander is a curve in a stream that results from erosion on the outside river bank and deposition on the inside river bank.__

Mouth __The mouth of a river is the end of the river where it meets the sea.__

Tributary __A tributary is a smaller stream that flows into a larger one.__

V-shaped valley __A V-shaped valley is a valley with steep sides produced by erosion from running water.__

Running water: Earth's most important agent of erosion.

Chapter 26: Weather
Layers of the Atmosphere

Label each of the layers of the atmosphere in the diagram (not drawn to scale).

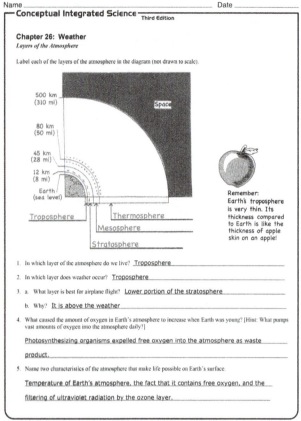

500 km
(310 mi)

Space

80 km
(50 mi)

45 km
(28 mi)

12 km
(8 mi)

Earth
(sea level)

Troposphere

Thermosphere

Mesosphere

Stratosphere

Remember:
Earth's troposphere is very thin. Its thickness compared to Earth is like the thickness of apple skin on an apple!

1. In which layer of the atmosphere do we live? __Troposphere__

2. In which layer does weather occur? __Troposphere__

3. a. What layer is best for airplane flight? __Lower portion of the stratosphere__

   b. Why? __It is above the weather__

4. What caused the amount of oxygen in Earth's atmosphere to increase when Earth was young? [Hint: What pumps vast amounts of oxygen into the atmosphere daily?]

   __Photosynthesizing organisms expelled free oxygen into the atmosphere as waste__
   __product.__

5. Name two characteristics of the atmosphere that make life possible on Earth's surface.

   __Temperature of Earth's atmosphere, the fact that it contains free oxygen, and the__
   __filtering of ultraviolet radiation by the ozone layer.__

Chapter 26: Weather
Layers of the Atmosphere—continued

6. a. What is the ozone layer?

   __A region with a high concentration of ozone molecules, O₃__

   b. Where is it located? __In the stratosphere__

   c. Mark it on the diagram with a dashed line.

   d. What does the ozone layer do for you? __It absorbs ultraviolet radiation, preventing__
   __harmful amounts of UV radiation from reaching Earth's surface. UV radiation has__
   __harmful effects ranging from skin and eye damage to cancer and disruption of DNA.__

7. What is the ionosphere?

   __A region containing a high concentration of ions—atoms or molecules with an__
   __electric charge__

   b. Where is it located? __In the upper mesosphere and lower thermosphere__

   c. Mark it on the diagram with a dotted line.

8. Does temperature increase or decrease with altitude in the troposphere? __Decrease__

   b. What is the reason?

   __The density of heat absorbing atmospheric gas molecules decreases with__
   __distance.__

9. What is wrong with the picture of the layers of the atmosphere? [Hint: Look at the scale. Can the layers of the atmosphere be drawn to scale on a diagram of this size?]

   __The diagram is not to scale. For instance, the picture represents the atmosphere as__
   __being much thicker than it really is compared with Earth—Earth's atmosphere is really__
   __only as thick as the skin on an apple. Also, looking at the vertical scale of the__
   __diagram we see that the troposphere is only about 1/150 of the thickness of Earth's__
   __entire atmosphere. So if the troposphere were drawn large enough to see, the entire__
   __atmosphere would not fit on a page of this book.__

Chapter 26: Weather
Prevailing Winds—How Do They Work?

Fill in the blanks to show where the following prevailing winds or "global winds" are located:
Polar easterlies, Westerlies; Northeast tradewinds; Southeast tradewinds.

These winds are the __polar easterlies__

These winds are the __westerlies__

These winds are the __NE tradewinds__

Is pressure here high or low? __high__

Is pressure here high or low? __low__

Doldrums

These winds are the __SE tradewinds__

These winds are the __westerlies__

These winds are the __polar easterlies__

N
60 N
30 N
0
30 S
60 S
S

Circle the correct answers:

1. Global winds are part of a huge pattern of (water) **(air)** circulation around the globe.

2. Global winds transport heat around the globe, which affects **(temperature)** (solar energy).

3. There are (two) (four) **(six)** wind belts on Earth. Within each belt, warm air rises, moves laterally, then sinks. Wind belts are **(convection cells)** (zones of high pressure).

4. The turning or spiraling of global winds is due to Earth's (tilt) **(rotation)**.

5. Like all winds, global winds flow from areas of **(high)** (low) pressure to regions of (high) **(low)** pressure.

6. Earth's atmosphere has regions of high and low pressure because of uneven **(heating)** (rotating) of Earth.

7. Global winds drive ocean **(currents)** (waves)—circulating streams of water that redistribute heat throughout the globe.

8. Warm air rises at the (poles) **(equator)** and sinks at the **(poles)** (equator).

Chapter 26: Weather
Read a Weather Map

Weather Symbols

To report the weather in shorthand, weather symbols are used. Read the weather symbols on the chart below.

Weather Map Symbols

Weather Conditions	Cloud Cover	Wind Speed (mph)	Special Symbols
Light Rain	No Clouds	Calm	Cold Front
Moderate Rain	One-Tenth or Less	3–8	Warm Front
Heavy Rain	Two- to Three-Tenths	9–14	H High Pressure
Drizzle	Broken	15–20	L Low Pressure
Light Snow	Nine-Tenths	21–25	Hurricane
Moderate Snow	Overcast	32–37	
Thunderstorm	Sky Covered	44–48	
Freezing Rain		55–60	
Haze		66–71	
Fog			

Use the above chart to answer the following questions:

1. a. What is the symbol for light snow? __✳ ✳__

   b. What is the symbol for a sky covered by clouds? __⊗__

   c. What is the symbol for winds between 55 and 60 miles per hour? _____

Conceptual Integrated Science — Third Edition

Chapter 26: Weather
Read a Weather Map—continued

Station Models

Weather data collected at weather stations can be represented on a weather map. Each weather station uses a station model—a summary of the weather that is based on weather symbols. Notice that the station model for Portland shows the city's weather at a certain time.

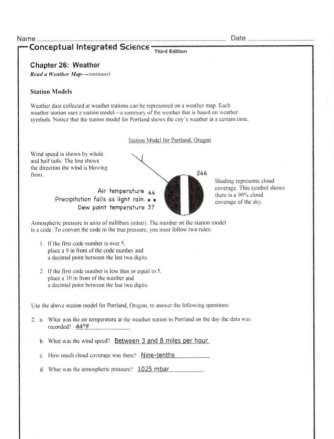

Station Model for Portland, Oregon

Wind speed is shown by whole and half tails. The line shows the direction the wind is blowing from.

Air temperature 44
Precipitation falls as light rain. ••
Dew point temperature 37

246

Shading represents cloud coverage. This symbol shows there is a 90% cloud coverage of the sky.

Atmospheric pressure in units of millibars (mbar). The number on the station model is a code. To convert the code to the true pressure, you must follow two rules:

1. If the first code number is over 5, place a 9 in front of the code number and a decimal point between the last two digits.

2. If the first code number is less than or equal to 5, place a 10 in front of the number and a decimal point between the last two digits.

Use the above station model for Portland, Oregon, to answer the following questions:

2. a. What was the air temperature at the weather station in Portland on the day the data was recorded?   44°F

   b. What was the wind speed?   Between 3 and 8 miles per hour

   c. How much cloud coverage was there?   Nine-tenths

   d. What was the atmospheric pressure?   1025 mbar

Conceptual Integrated Science — Third Edition

Chapter 26: Weather
Read a Weather Map—continued

Weather Maps

1. Describe the weather in each of these cities. Include all the information on each station model.

   a. Sioux Falls

   29°F; westerly wind blowing between 3 and 8 miles per hour; atmospheric pressure of 1029.6; no precipitation

   b. Amarillo

   46°F; northerly wind blowing between 21 and 25 miles per hour; atmospheric pressure of 965.6; complete cloud coverage; thunderstorms

   c. New York City

   41°F; southwesterly wind blowing between 15 and 20 miles per hour; atmospheric pressure of 1010.8; overcast; light rain

---

Conceptual Integrated Science — Third Edition

Chapter 26: Weather
Read a Weather Map—continued

2. The arrows on the cold front symbol show the cold front is moving and in what direction. What is the direction in which the cold front is moving?

   The cold front is moving east.

3. How is the cold front affecting the weather? (Compare the regions ahead of the front with the locations the front has recently passed through.)

   The cold front brings westerly winds. It is blowing from a region of higher pressure to lower pressure. Thunderstorms are at the front in Amarillo. The sky is cloudy in regions the cold front is approaching. Clouds are ahead of the front while skies are clearing behind it.

Whenever you lose your way, run slower—not faster!

For beginners, wisdom is knowing not to throw a rock straight up. Advanced wisdom is knowing what to overlook.

Conceptual Integrated Science Third Edition

Chapter 27: Environmental Geology
Earthquakes: How Big and How Often?

Earthquake Magnitude and Frequency

Richter Scale Magnitude	Earthquake Effects	Average Number Per Year
less than 2.0	Cannot be felt	600,000+
2.0 to 2.9	Recorded but cannot be felt	300,000
3.0 to 3.9	Felt by most people near epicenter	49,000
4.0 to 4.9	Minor shock; slight damage near epicenter	6,000
5.0 to 5.9	Moderate shock; energy released equals the energy released by one atomic bomb	1,000
6.0 to 6.9	Large shock; damaging to population centers	120
7.0 to 7.9	Major earthquake with severe property damage; can be detected around the world	14
8.0 to 8.9	Great earthquake; communities near epicenter are destroyed. Energy released is equivalent to that of millions of atomic bombs.	once every 5 to 10 years
9.0 to 9.9	Large earthquake recorded	1 to 2 per century

1. What does the Richter scale measure?

It is a measurement of how much the ground shakes during a quake.

2. About how many earthquakes occur each year but are not felt by people?

At least 600,000

3. How many earthquakes of magnitude 5.0 to 5.9 does it take to release the same amount of energy released by one 8.0 to 8.9 earthquake?

Millions of 5.0 to 5.9 earthquakes release the energy equivalent of one 8.0 to 8.9 earthquake.

Conceptual Integrated Science Third Edition

Chapter 27: Environmental Geology
Earthquakes: How Big and How Often?—continued

4. a. According to the chart on the previous page, what is a "major earthquake"?

Major earthquakes cause major damage to buildings and other property and can be detected around the world.

b. What is the range of Richter magnitudes of a major earthquake?

Magnitudes between 7.0 and 7.9 on the Richter scale

c. How many earthquakes occur each year?

Fourteen major earthquakes on the average

5. Where do most of the world's earthquakes occur?

In the region surrounding the Pacific Ocean known as the Ring of Fire

6. Suppose you are near an earthquake epicenter. You can feel the quake but there is no damage. What is the magnitude of the earthquake?

Most likely has a magnitude between 3.0 and 3.9

7. a. Would you rather endure ten magnitude 3.0 earthquakes or one 8.0 earthquake?

Endure ten magnitude 3.0 earthquakes

b. Why?

You and your belongings are less likely to be injured in many small earthquakes than in one large one. The force of ground shaking exists during the quake but then is over, so the effects from small earthquakes don't accumulate.

> Earth quakes, shakes, rocks, and rolls—what a ride!

> The process called science replaces *confusion* with *understanding* in a manner that's precise, predictive, and reliable — while providing an empowering and emotional experience.

Conceptual Integrated Science Third Edition

Chapter 27: Environmental Geology
Is Your Knowledge Shaky?

Use the clues below to fill in the words of the crossword puzzle in the numbered blanks on the facing page.

Across

1. Most earthquakes occur at   plate boundaries   .

5. Force that builds up in rock prior to an earthquake

6. What is released in the form of waves in an earthquake?

9. Seismic   waves are triggered by rocks slipping at an earthquake focus.

11. The Ring of Fire   is a region surrounded by convergent plate boundaries where frequent earthquakes occur.

Down

2. Point at Earth's surface directly above an earthquake focus.

3. The place inside Earth where rock slips, starting an earthquake.

4. A crack in the crust where blocks of rock have moved relative to one another.

7. The earthquake scale that measures shaking of the ground.

8. An instrument that measures earthquake magnitude on the Richter scale.

10. What force holds blocks of strained rock together before an earthquake?

> I like puzzles.

Chapter 27: Environmental Geology
Is Your Knowledge Shaky?—continued

Who was the Richter Scale named after?

Maybe Charles Scale?

Chapter 27: Environmental Geology
Volcano Varieties

Volcanoes are mountains built up of erupted rocky debris, including lava, ash, and rocky fragments called *pyroclastics*. Each drawing here represents one of the three types of volcanoes—shield, cinder cone, and composite. Shield volcanoes, cinder cones, and composite volcanoes have different sizes and shapes.

1. Label the type of volcano each drawing represents.

Type of volcano is   shield volcano
F   H   I   K
....

Type of volcano is   composite cone
A   C
D   E
J   L

Type of volcano is   cinder cone
B   G
....

2. On the dashed lines above, write the letter (A to L) that is appropriate for the type of volcano.

A. Also called a "stratovolcano"   B. Small but steep volcanoes   C. Tall, broad volcanoes

D. Built from layers of lava,   E. Can trigger *lahars*   F. Kilauea is an example
   ash, and pyroclastics

G. Built up from ash and   H. Gently sloping volcanoes   I. Built from cooled, runny
   cinders   lava

J. Mt. St. Helens is an example   K. Usually erupts quietly   L. Often erupt explosively

---

Chapter 27: Environmental Geology
Atmospheric Carbon Dioxide and Global Temperature: A Match?

Scientists have been able to estimate Earth's average temperature as well as the concentration of carbon dioxide in Earth's atmosphere over the past several hundred thousand years. The graph below shows this.

Temperature and Carbon Dioxide Concentration Over 160,000 Years

KEY
/ Carbon dioxide concentration
/ Average global temperature

Can you interpret the graph to answer these questions?
1. Over the past 160,000 years, when was Earth warmest?

   Earth's average temperature was greatest about 135,000 years ago.

2. When Earth reached its highest average global temperature, was atmospheric carbon dioxide also very high?
   Yes

3. Do the graphs show a *correlation* (connection) between atmospheric carbon dioxide and Earth's average temperature?

   Yes, the graphs have the same basic pattern so the graphs do correlate with one

   another.

Chapter 27: Environmental Geology
Atmospheric Carbon Dioxide and Global Temperature: A Match?—continued

4. a. Were all the variations in atmospheric carbon dioxide caused by human activities?

      No

   b. What is the reason for your answer?

      Carbon dioxide levels reached minima and maxima in the Pleistocene era.

      The human population was small and culture nonexistent until

      the Holocene era, which began about 10,000 years ago. A principal natural cause

      of high levels of carbon dioxide is the eruption of volcanoes.

Scientists have carefully collected data on atmospheric carbon dioxide and temperature for the past 1,000 years. This data is represented on the graph below.

Temperature and Carbon Dioxide Concentration Over 1,000 Years

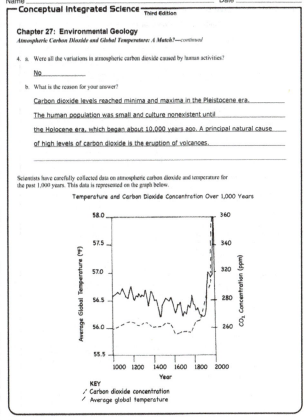

KEY
/ Carbon dioxide concentration
/ Average global temperature

331

Conceptual Integrated Science — Third Edition

Chapter 27: Environmental Geology

Atmospheric Carbon Dioxide and Global Temperature: A Match?—*continued*

Use the graph of temperature and carbon dioxide concentration over 1,000 years on the facing page to answer the questions.

5. a. What was Earth's average temperature in 1000 C.E.? About 56.5°

    b. What was it in 1900? 56.2°F

    c. In the year 2000? 58°F

6. Do the graphs show a *correlation* between atmospheric carbon dioxide and Earth's average temperature since 1860?

Yes. The graphs almost overlap, showing a close correlation.

7. Why did carbon dioxide levels in the atmosphere begin to rise dramatically around 1860?

The industrial revolution began in the mid-nineteenth century. It marked the

beginning of the era of widespread fossil fuel use.

8. a. Does the graph show that increased levels of carbon dioxide in the atmosphere *cause* global warming?

No

    b. Explain.

Graphs show correlation, not causation. Yet, the high degree of correlation,

coupled with our understanding of the mechanism by which greenhouses retain

terrestrial radiation and increase global temperature, strongly suggests a causal

relationship.

I'd rather hang out with friends who have reasonable doubts than ones who are absolutely certain about everything.

Conceptual Integrated Science — Third Edition

Update the Law on Climate Change

Congratulations, you have won the election! You are now a local, state, or federal lawmaker. Your first priority is to draft a law on climate change. Your law can specify a policy change to either mitigate climate change or help the public adapt to it. You will need to follow the process below to draft your new law and get it passed.

1. Do Your Research

Review an environmental law to get a sense of how laws are written. Research online or through your local library or town hall to find an environmental law, or *statute*, that you can understand. Note that environmental laws generally include the following:
*   Statement of why the law is needed
*   Definition of terms
*   Statement of who is affected and who is exempt
*   Description of how the law will be enforced
*   Statement of penalties to any parties violating the law
*   Authorization of funding

2. Formulate Your Policy Idea

What kinds of laws are needed in an era of changing climate? Formulate your idea for a new law. Write your statement of intent below. That is, describe the proposed policy change and explain how it will help society mitigate or adapt to climate change. (Include blank lines for a short paragraph.)

3. Define the Scale of Change

Is the change you are proposing a local one that would best be handled by the local authorities? Or does your law involve state-level change, so it needs to be implemented by the state legislature? Or is your law national in scope? If so, it is *federal* legislation that must be handled at the federal level—by Congress.
    Decide if you are playing the role of a local, state, or federal lawmaker and write it in the blank below. (Include blank for a few words.)

4. Be Practical

Now that you have a basic idea for your law and its scale, refine it. Review the following practical considerations. Write your answers to these questions below.
a) Does your law have loopholes? How can you close them?
b) How will your law affect jobs and the economy?
c) How much will it cost to implement your law? Where will the money come from?
d) Is the law fair? Is it biased to favor particular groups? Does it impact any demographic group in an overly burdensome manner? (Include a couple of blank lines for parts a–d.)
e) Cite relevant scientific sources to show that the change you propose is feasible in terms of available technology. Also include an environmental impact statement. (Gathering this information for actual legislation would be an exhaustive practice. For the purposes of this assignment, you can limit yourself to 3–5 scientific sources.)

5. Draft Your Law

It's time to roll up your sleeves and draft your law. Write it on a separate piece of paper. It should be detailed and include provisions to make it doable, fair, and effective. (Include blank lines for a short paragraph.)

6. Assess Your Impact

Work with a partner. Let your partner read your law while you read theirs. Write an assessment of the law your partner drafted. Include the following points: 1) What is the most constructive aspect of the law? 2) What could be the short-term impacts of the law? 3) What would be the long-term impacts? 4) Would you vote for or against the law as it is currently written? 5) How could the law be amended so you would vote for it?

7. Maximum Impact: Work On Federal Legislation

Work in small groups of four to six students. Choose group members so that at least one group member has proposed a new federal law. Your group will model the process of taking proposed legislation from committee to be voted on by the full U.S. Senate or House of Representatives. Your group will play the role of a Senate or Congressional committee. First step: Review the legislative process described in the box below.

Conceptual Integrated Science — Third Edition

> *To Change the Law: Debate and Persuade!*
>
> Before a proposed federal law or *bill* is passed into law, it must be debated by politicians representing citizens all across the nation. Therefore, the bill will be debated by representatives of opposing political parties who have different political viewpoints but many common interests, values, and beliefs.
>   The bill begins in committee, a bipartisan subgroup of congress persons or senators. If it is approved in committee, the bill moves on to the full House of Representatives or Senate. If the bill wins the vote in both the House of Representatives and the Senate, it becomes law.

    Now, allow each member of your group who has authored federal legislation to present their law to the committee. The author reads the bill aloud and explains key features and benefits. Appoint one group member to take notes to keep a record of the debate.

8. Make Amendments

Committee members take turns responding to the proposed bill. Your goal is for the committee to debate every important aspect of it. Meeting notes should be thorough.
    Allow committee members to propose amendments to each bill. Vote on proposed amendments to decide whether they should be included. Now have a final vote in your committee. Is the bill ready to submit to the full chamber?

9. The Final Vote

Your entire class will play the role of the full chamber. Select one member from each committee to present all the bills your committee has approved. Allow questions from the full chamber. Now vote!

10. Reflect

Choose one bill that was passed into law. Explain why that bill was successful in getting through the entire process.

Conceptual Integrated Science *Third Edition*

Chapter 28: The Solar System
Earth–Moon–Sun Alignment

Here we see a shadow on a wall cast by an apple. Note how the rays define the darkest part of the shadow, the *umbra*, and the lighter part of the shadow, the *penumbra*. The shadows that comprise eclipses of planetary bodies are similarly formed. Below is a diagram of the Sun, Earth, and the orbital path of the Moon (dashed circle). One position of the Moon is shown. Draw the Moon in the appropriate positions on the dashed circle to represent (a) a quarter moon, (b) a half moon, (c) a solar eclipse, and (d) a lunar eclipse. Label your positions. For c and d, extend rays from the top and bottom of the Sun to show umbra and penumbra regions.

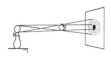

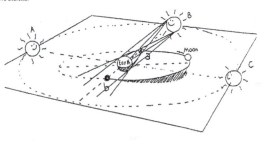

The diagram below shows three positions of the Sun: A, B, and C. Sketch the appropriate positions of the Moon in its orbit about Earth for (a) a solar eclipse and (b) a lunar eclipse. Label your positions. Sketch solar rays similar to the above exercise.

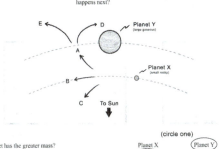

Conceptual Integrated Science *Third Edition*

Chapter 28: The Solar System
Pinhole Image Formation

Look carefully at the round spots of light on the shady ground beneath trees. These are *sunballs*, which are images of the Sun. They are cast by openings between leaves in the trees that act as pinholes. (Did you make a pinhole "camera" back in middle school?) Large sunballs, several centimeters in diameter or so, are cast by openings that are relatively high above the ground, while small ones are produced by closer "pinholes." The interesting point is that the ratio of the diameter of the sunball to its distance from the pinhole is the same as the ratio of the Sun's

diameter to its distance from the pinhole. We know the Sun is approximately 150,000,000 km from the pinhole, so careful measurements of the ratio of diameter/distance for a sunball leads you to the diameter of the Sun. That's what this page is about. Instead of measuring sunballs under the shade of trees on a sunny day, make your own easier-to-measure sunball.

150,000,000 km

1. Poke a small hole in a piece of card. Perhaps an index card will do, and poke the hole with a sharp pencil or pen. Hold the card in the sunlight and note the circular image that is cast. This is an image of the Sun. Note that its size doesn't depend on the size of the hole in the card, but only on its distance. The image is a circle when cast on a surface perpendicular to the rays—otherwise it's "stretched out" as an ellipse.

2. Try holes of various shapes, say, a square hole or a triangular hole. What is the shape of the image when its distance from the card is large compared with the size of the hole? Does the shape of the pinhole make a difference?

    <u>THE IMAGE IS ALWAYS A CIRCLE. THE SHAPE OF THE PINHOLE IS *NOT* THE SHAPE</u>
    <u>OF THE IMAGE CAST THROUGH.</u>

3. Measure the diameter of a small coin. Then place the coin on a viewing area that is perpendicular to the Sun's rays. Position the card so the image of the sunball exactly covers the coin. Carefully measure the distance between the coin and the small hole in the card. Complete the following:

    $$\frac{\text{Diameter of sunball}}{\text{Distance to pinhole}} = \frac{d}{h} = \frac{1}{110} \left(\text{SO SUN'S DIAM} = \frac{1}{110} \times 150,000,000 \text{ km} \right)$$

    With this ratio, estimate the diameter of the Sun. Show your work on a separate piece of paper.

4. If you did this on a day when the Sun was partially eclipsed, what shape of image would you expect to see?

    <u>UPSIDE-DOWN CRESCENT, IMAGE OF THE PARTIALLY</u>
    <u>ECLIPSED SUN</u>

WHAT SHAPE DO SUNBALLS HAVE DURING A PARTIAL ECLIPSE OF THE SUN?

Conceptual Integrated Science *Third Edition*

Chapter 28: The Solar System
Jumping Jupiter

Planet X approaches Planet Y in close proximity. What happens next?

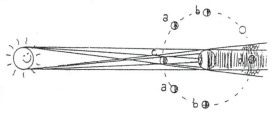

1. Which planet has the greater mass?

2. Which planet has the greater orbital speed?

                                                          (circle one)

Planet X   (Planet Y)

(Planet X)   Planet Y

3. Five potential paths (A, B, C, D, E) for planet X are indicated with curved arrows. Describe what makes each path possible or not possible.

    A: <u>Planet X is "pulled upward" by its gravitational attraction to planet Y.</u>

    B: <u>If planet X is close to planet Y, then maintaining path B is not likely.</u>

    C: <u>Given planet X's momentum and attraction to planet Y, path C is not possible.</u>

    D: <u>If planet X is sufficiently close to Y, the gravity may be strong enough to yield path D.</u>

    E: <u>Being close enough for path A, but not close enough for path D, planet X with its</u>
    <u>momentum may swing outward to path E.</u>

4. How likely is it for planet X to become a moon with a stable orbit upon following path D?

    A stable orbit is possible but conditions would have to be exactly right. More likely is a case where planet X eventually collides with planet Y or is thrown radically off course.

5. What happens to the orbital velocity of planet X upon following path E?

    Initially, the orbital velocity slows down as planet X is now farther from the Sun. The shape of its orbit, however, may now be more "elliptical," which means a greater variablity in its orbital velocity: faster when closer to the Sun and slower when farther from the Sun.

6. What happens to the distance between the Sun and planet Y upon planet X following path E?

    If planet X is pulled upward, then planet Y is necessarily pulled downward, closer to the Sun, although only by a little because of its much greater mass.

7. What happens to the orbital speed of planet Y upon planet X following path E?

    Initially, the orbital velocity speeds up as planet Y is now closer to the Sun. The shape of its orbit, however, may now be more "elliptical," which means a greater variablity in its orbital velocity: faster when closer to the Sun and slower when farther from the Sun.

See the Conceptual Academy video tutorial on planet Neptune for further discussions.

Conceptual Integrated Science *Third Edition*

Chapter 28: The Solar System
Rings of Saturn

Two small rocks, A and B, are in orbit around a huge gaseous planet.

1. Gravity from the planet is weaker for which rock:

                    (circle one)

                    A   (B)

2. Which must move sideways faster to remain in orbit:

                    (A)   B

3. What eventually happens to the distance between them?   <u>It starts to increase.</u>

A moon made of soft clay is in orbit. Draw arrows to indicate the orbital velocities of the near and far sides of this soft moon.

4. Why does the shape of this moon elongate?   <u>The front needs to move faster than the back.</u>

5. Might this moon eventually rip apart?   <u>Yes</u>

6. If this moon were instead made of iron, electrical forces of attraction between iron atoms would minimize the elongation. But there's another reason elongation would be minimized. What is it?

<u>The reason is the attractive force of gravity, which is why large freely floating bodies in space tend</u>
<u>to be round.</u>
<u>If one part stuck upward, gravity would tend to pull it right back down.</u>

Consider the planetary ring system shown to the left where each ring consists of countless small chunks of water ice.

7. Rank the rings in order of increasing orbital velocity

    <u>A</u> > <u>B</u> > <u>C</u>

8. Do you suppose the ice chunks within a single ring are each orbiting at the same velocity?

    No!

9. Why can't these particles coalesce together into a moon?

    They keep moving relative to each other.

10. What would be the fate of Saturn's rings if each particle were made of rock rather than ice?

Rock is more dense than ice. This means more mass, which means more gravity, thus allowing the them to start clumping, perhaps into a moon.

See the Conceptual Academy video tutorial on planet Saturn for further discussions.

Conceptual Integrated Science
Third Edition

Chapter 28: The Solar System
Word Play

After you have read Chapter 28, use this word play to help solidify some key terms. Only refer back to the textbook after you have given this word play a solid try. Remember, the more you attempt to articulate what you think you understand, the greater the durability of that understanding.

1. A type of nebula having something in common with our atmosphere
2. A prominent constellation located along the celestial equator
3. A slowly rotating cloud of these likely formed our solar system
4. Formed upon the formation of a star
5. The growth of a massive object by gravitational attraction of matter is called
6. A dense interstellar cloud that obscures light
7. Type of red nebula containing hydrogen atoms ionized by nearby stars
8. An intra-galactic cloud of gas and dust

1.   r e (f) l e c t i o n
2.   o r (i) o n
3. p l a n e t e (s) i m a l s
4.   (h) e l i u m
5.   a c c r e (t) i o n
6.   d (a) r k
7.   e m (i) s s i o n
8.   n e b u (l) a

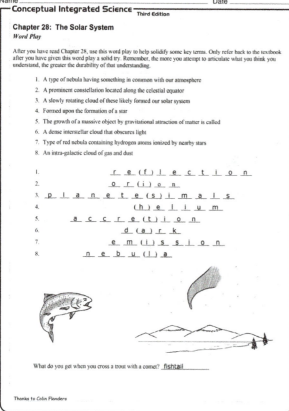

What do you get when you cross a trout with a comet? __fishtail__

Thanks to Colin Flanders

Conceptual Integrated Science
Third Edition

Chapter 28: The Solar System
Solar Anatomy

After you have studied Chapter 28, identify the following structures of our Sun. Try your best to remember these structures before looking back to the chapter for confirmation.

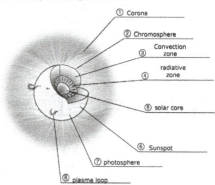

① Corona
② Chromosphere
③ Convection zone
④ radiative zone
⑤ solar core
⑥ Sunspot
⑦ photosphere
⑧ plasma loop

Once you are proficient at recalling the names of the above structures, write down your understanding, in brief, of what happens in structures 3, 4, 5, and 8.

3. Heat is transferred from the bottom of the convection zone to the top primarily by the rising of bulk units of plasma. This can happen because a parcel of plasma at the bottom is able to expand in volume, making it less dense, allowing it to rise.

4. Plasma within the radiation zone is too dense to allow for the expansion of parcels. Hence, no convection. Instead, energy travels outward primarily by the release of photons (radiation), which is relatively slow as photons move both outward and inward.

5. The solar core in our Sun is where nuclear fusion is primarily taking place. Interestingly, the energy output per cubic meter is relatively small (comparable to a compost heap). But the solar core is so large that the total amount of energy is astronomical.

8. Movement of plasma within the Sun generates intense magnetic fields. Because of its fluidity, plasma along the equator of the Sun rotates faster than at the pole. Magnetic field lines get twisted into loops through which plasma travels.

Conceptual Integrated Science
Third Edition

Chapter 28: The Solar System
Distance to the Moon

It takes the Moon 27.3 days to orbit Earth. We can use this observation to calculate the distance to the Moon. First, convert 27.3 days into seconds using this string of conversion factors:

from the days in a month!!

$$(27.3 \text{ days})\left(\frac{24 \text{ hours}}{1 \text{ days}}\right)\left(\frac{60 \text{ min}}{1 \text{ hours}}\right)\left(\frac{60 \text{ sec}}{1 \text{ min}}\right) = 2{,}358{,}720 \text{ sec}$$

In this many seconds, the Moon travels a circumference, which equals $2\pi r$, where "r" is the radius, which is the distance between the centers of Earth and the Moon. This distance traveled over seconds is the Moon's orbital velocity:

$$\frac{\text{distance}}{\text{time}} = \frac{\text{circumference}}{\text{time}} = \left(\frac{2\pi r}{2{,}358{,}720}\right) = V_{orbit}$$

enter seconds here →   orbital velocity

Orbital velocity is also given by the equation: $\left[V_{orbit}\right]^2 = \dfrac{GM_e}{r}$

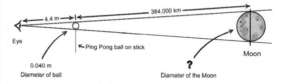

radius, r   Moon

Earth

Where G is the gravitational universal constant, $6.6677 \times 10^{-11} \frac{\text{N} \cdot \text{m}^2}{\text{kg}^2}$, and M is the mass of Earth, 5.97×10^{27} kg.

Combine the above two equations to calculate the distance to the Moon, r. Your answer should be around 383,000 km.

$$\left[\frac{2\pi r}{2{,}358{,}720}\right]^2 = \frac{GM_e}{r}$$

enter seconds here

Use this space to solve for r

Start by bringing "r" to one side

$$\frac{(2\pi r)^2}{(2{,}358{,}720)^2} = \frac{GM_e}{r}$$

$$(2\pi r)^2 r = GM_e (2{,}358{,}720)^2$$

$$4\pi^2 r^3 = (6.677 \times 10^{-11} \tfrac{N \cdot m^2}{kg^2})(5.97 \times 10^{24} kg)(2{,}358{,}720 m)^2$$

$$r^3 = \frac{2.22 \times 10^{26} \, m^3}{4(3.1415)^2} = 5.62 \times 10^{25} \, m^3$$

Math Counts!

Cubed root is the same $(y)^{0.3333}$

$$r = \sqrt[3]{5.62 \times 10^{25} m^3}$$

"close" to accepted value!

$$r = 383{,}000{,}000 \, m \Rightarrow 383{,}000 \, km$$

Conceptual Integrated Science
Third Edition

Chapter 28: The Solar System
Diameter of the Moon

Hold a Ping Pong ball far out enough so that it appears to be the same size of the Moon. Note how the phase of the ball and the Moon are the same.

The distance between you and the Ping Pong ball will be about 4.4 meters. The Ping Pong ball itself is 40 centimeters wide, which is 0.040 meters.

     4.4 m   384,000 km

Eye   ← Ping Pong ball on stick   Moon

0.040 m
Diameter of ball   **?**
Diameter of the Moon

From the previous page we know that the distance to the Moon is about 384,000 km. Use all this information to calculate the diameter of the Moon. Use the following ratio:

$$\frac{4.4 \text{ m}}{0.040 \text{ m}} = \frac{384{,}000 \text{ km}}{?}$$

← Diameter of the Moon in kilometers

Use this area to show your work:

cross multiply

$$(4.4 \, m)(?) = (384{,}000 \, km)(0.040 \, m)$$

$$? = \frac{(384{,}000 \, km)(0.040 \, m)}{(4.4 \, m)}$$

$$= 3{,}500 \text{ km}$$

Conceptual Integrated Science — Third Edition

Chapter 29: The Universe
Stellar Parallax

Finding distances to objects beyond the solar system is based on the simple phenomenon of *parallax*. Hold a pencil at arm's length and view it against a distant background—each eye sees a different view (try it and see). The displaced view indicates distance. Likewise, when Earth travels around the Sun each year, the position of relatively nearby stars shifts slightly relative to the background stars. By carefully measuring this shift, astronomer types can determine the distance to nearby stars.

> Can you see why the close star appears to shift positions relative to the background stars? And how maximum shift appears in observations 6 months apart?

Background stars
Close star
Angle of parallax shift
Earth

The photographs below show the same section of the evening sky taken at a 6-month interval. Investigate the photos carefully and determine which star appears in a different position relative to the others. Circle the star that shows a parallax shift.

A   B

Below are three sets of photographs, all taken at 6-month intervals. Circle the stars that show a parallax shift in each of the photos.

Set A   Set B   Set C

Use a fine ruler and measure the distance of shift in millimeters and place the values below:

Set A __19__ mm   Set B __6.3__ mm   Set C __12.6__ mm

Which set of photos indicates the closest star? The most distant "parallaxed" star?
The star that shows the greatest shift in position relative to background stars is the star closest to us. The circle star in Set A is thus closest, while the circled star in Set B is farthest. B

Conceptual Integrated Science — Third Edition

Chapter 29: The Universe
Black Holes

Imagine a ship in orbit around a black hole. An onboard clock reads 1:30 as does the wrist watch of an astronaut about to be lowered toward the black hole.

1:30

1. As the astronaut is lowered toward the black hole, what happens to the gravitational force between the astronaut and the black hole?

   The force increases

2. How does this impact the spaceship?

   Attached by the tether there's a downward pull.

3. What must the spaceship do to maintain orbit?

   Tough question! One thing the ship could do is seek a higher orbit in order to maintain the same center of gravity.

Photon sphere
Event horizon
Black hole

Astronaut's Point of View

4. As the astronaut passes through the photon sphere, what does she notice about the time on her watch?

   The time on her watch passes as normally as ever, 60 seconds in each minute.

5. How about when she passes through the event horizon?

   The time on her watch passes as normally as ever, 60 seconds in each minute.

6. As the astronaut passes through the photon sphere, what does she notice about the time on the ship's clock?

   The time on the ship's clock is passing by unusually fast. For example, while she experiences 60 seconds, the clock on the ship might show an hour passing.

7. How about when she passes through the event horizon?

   The time on the ship's clock passes through infinity. Of course, by then the ship is gone.

Ship's Point of View

8. Assume the astronaut experienced 10 minutes as she was lowered to the photon sphere. How many minutes did her shipmates find it took her to be lowered that far? (circle one)

   a) less than 10 minutes   b) 10 minutes   c) more than 10 minutes ⟵ circled

---

Conceptual Integrated Science — Third Edition

??
??

What time you would place on these clocks depends upon your point of view. Within each point of view the clock seems to be running normally. It's only when you look at the other's clock that things get weird. The astronaut will see the ship's clock running fast. The ship's crew will see the astronaut's watch running slow. There would also be shifts in the color of light as is explored in the video tutorials.

Photon sphere
Event horizon
Black hole

Theories combining quantum mechanics with Einstein's general relativity suggest that the falling astronaut would encounter a shell of fire upon reaching the black hole's event horizon. If tidal forces upon entering the black hole didn't kill the astronaut, then, this "shell of fire" certainly would.

Ship's Point of View (continued)

9. Imagine the tether unexpectedly cuts loose just after the astronaut has passed into the photon sphere. From the ship's point of view, how long does it take for the falling astronaut to reach the event horizon?

   A really, really, really long time—like forever!

10. By the time the astronaut has passed through the event horizon, what has happened to our universe as we know it?

   Heat death, the Big Rip, external inflation—something like that. Basically, it will no longer resemble what it once was or it may, perhaps, no longer exist.

11. Is it possible for an object to pass through a black hole's event horizon?

   Yes. FYI: For a small black hole (100 solar masses) the tidal forces would likely rip the structure apart. For a super massive black hole (billions of solar masses), however, the tidal forces may be surviable for that object.

12. How long does this take?   (circle one)

   (a) From the object's point of view: not long   forever

   (b) From our point of view: not long   forever ⟵ circled

Conceptual Integrated Science — Third Edition

Chapter 29: The Universe
Hertzsprung–Russell Diagram

The Hertzsprung-Russell diagram plots stellar temperature (color) by luminosity. Each dot on this diagram represents a star.

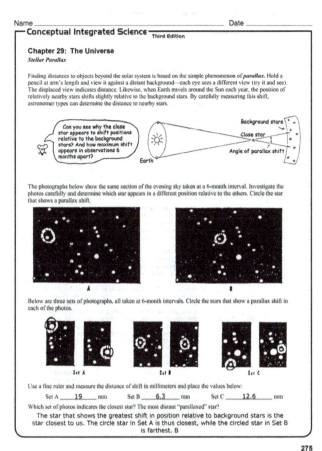

1. Draw an arrow pointing to the largest star on this diagram.
2. Draw a rectangle around the most quickly evolving stars.
3. Draw a triangle over the longest living star.
4. Circle the shortest lived stars.
5. Place an X over a white dwarf.
6. Trace a star shape over the approximate coordinates of our Sun.
7. Draw a large letter B over the bluish stars.
8. Draw a large letter R over the reddish stars.

9. How is the image from an H–Z diagram different from the image of an actual galaxy?

   The H–Z diagram does NOT show how the stars are related to each other within space. A red giant and a white dwarf shown vastly apart on this diagram might actually be within orbit around each other.

Shown below are H–Z diagrams from five different star clusters of various ages found within our Milky Way Galaxy.

8000 million years old   (b)   (c)   (d)   200 million years old

red giants
white dwarfs

10. Are these star clusters (left to right) presented from old to younger or from young to older? Please explain.

   These H–Z diagrams are placed from left to right from old to younger. The cluster represented on the far left is the oldest as evidenced by formation of red giants and numerous white dwarfs. So the H–Z diagram of a particular patch of stars, such as a galaxy? can tell us something about the age and character of that batch of stars.

Conceptual Integrated Science — Third Edition

Chapter 29: The Universe
Solar Life Cycle

1. Sort the following frames depicting a star's life cycle in the correct sequence:

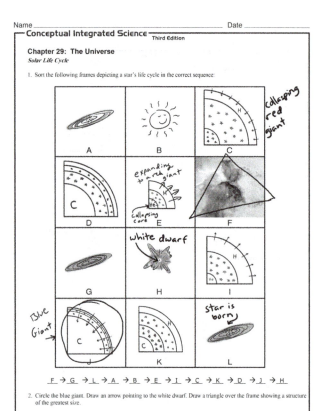

Labels on frames (handwritten):
- C: Collapsing red giant
- E: expanding to a red giant / collapsing core
- H: white dwarf
- J: Blue Giant
- L: star is born

$$\underline{F} \rightarrow \underline{G} \rightarrow \underline{L} \rightarrow \underline{A} \rightarrow \underline{B} \rightarrow \underline{E} \rightarrow \underline{I} \rightarrow \underline{C} \rightarrow \underline{K} \rightarrow \underline{D} \rightarrow \underline{J} \rightarrow \underline{H}$$

2. Circle the blue giant. Draw an arrow pointing to the white dwarf. Draw a triangle over the frame showing a structure of the greatest size.

Conceptual Integrated Science — Third Edition

Chapter 29: The Universe
The Big Picture

1. A chewed apple core lands next to an, who thinks, "How did this beautiful apple happen to land right next to me?" What do you tell the ant?

The ant doesn't undersatnd human language, but if it could, you might kindly suggest that this was a random incident and that recieving the apple core was pure luck.

2. A boy sees a license plate that reads CTX-4872. He ponders at the chances of seeing this particular plate on this particular day. What do you tell the boy?

There are billions of cars on the road with billions of license plates. This just happened to be the one he saw. However, if he lived in Vermont and saw a license plate from China, then that would be unusual.

3. A girl studying geography wonders whether it's a coincidence that almost all major cities are along a large body of water or a river. What do you tell the girl?

Humans need water to survive. It's no coincidence that early settlements were close to water and waterways

4. Johannes Kepler, the famous 17th century astronomer and mathemetician, tries to calculate why Earth is 150,000,000 kilometers from the Sun. What do you tell Kepler?

We now know that Earth just happens to be the distance it is from the Sun. However, if it were much closer or farther away, then conditions on this planet would not be so hospitable for life.

5. A cosmologist wonders why the repulsive (dark energy) and attractive (gravity) forces within our universe are nearly perfectly balanced. What might you suggest to the cosmologist?

Repulsive / Attractive

Perhaps if they were not so balanced, conditions would not be so hospitable to life and we would not be here to ponder such questions.

6. Many cosmologists now suspect there are other "universes" besides our own with different parameters and universal constants. If correct, what proportion of these "other universes" might be as stable as our own?

If the parameters of a universe are actually random, then the conditions within different universes would be wildly different. For example, they may be quite short lived as the repulsive or attractive forces win over causing either hyper-rapid expansion or hyper-rapid contraction.

7. Discuss with others what single word best fits all of these sentences:

An electron is a(n) __island__ within an atom.

Hawaii is a(n) __island__ within the Pacific Ocean.

The Earth is a(n) __island__ within our solar system.

Our solar system is a(n) __island__ within the galaxy.

Our galaxy is a(n) __island__ of stars within the universe.

Our universe is a(n) __island__ of galaxies within a larger multiverse.

Something like "speck" would be good answer too!